DRAGONS FROM THE DUNES

DRAGONS FROM THE DUNES

The Search for Dinosaurs in the Gobi Desert

J.R. LAVAS

Acknowledgements

Particular thanks to Zofia Kielan-Jaworowska, former head of Vertebrate Palaeontology at the Polish Palaeobiological Institute in Warsaw (now at the University of Oslo), who kindly reviewed sections of the manuscript, assisted in obtaining photos from other sources, and gave encouragement for both this project and for an archosaur bibliography completed earlier. Sergei Kurzanov of the Russian Palaeontological Museum in Moscow was most helpful in providing material from the archives of Anatole Rozhdestvensky, as well as photos from museum exhibits and recent Russian field work. Both authors supplied reprints of their publications. Wojciech Skarzynski (Polish Palaeobiological Institute) sent slides from the Polish-Mongolian expeditions and the Dinosaur Park at Chorzow, while Carmen Collazo (American Museum of Natural History, New York) tracked down specific negatives from the archives of the Central Asiatic Expeditions. Discussions and advice from Kendall Clements (Sydney University) was most appreciated, as was access to his reprint collection on dinosaur ecology and bird evolution. The manuscript was considerably improved by Philip Hair, Emil Melnichenko, and Mike Bradstock, while the pleasing format is due to Chris O'Brien who accommodated many revisions and gave advice throughout. Printing and binding was much facilitated by assistance from Ivan Mrsic and Ron Lietz. Thanks also to Richard & Louise Lavas, Andrew Fitton, Ruth Watson, Lynn Huddleston, Malcolm Ferguson, and the staff of David Bateman publishers.

All osteological drawings and some life reconstructions were redrawn from originals and at times modified in the light of recent evidence. Reasonable effort has been made to track original authors and artists, the credits for which are given in captions (authors describing original material by surname, others, including artists, with 1st initial also). The illustrations on pages iii,3 (both),4,5,38,45 (right),68,85 & 134 are based on drawings or photos in Efremov I.A., 1956, Road of the Wind, Trudrezervizdat, Moscow. All uncredited illustrations are covered by the book's copyright. Unless noted, some opinions such as those on dinosaur ecology, may be specifically my own and need not imply endorsement by the persons acknowledged above. I also take full responsibility for any errors, omissions, or misinterpretations in the text or captions.

Finally, no book of this nature can be complete without reference to the late Zdenek Burian who, more than anyone, so convincingly captured the prehistoric world on canvas. In hundreds of evocative reconstructions in palaeontology and anthropology, Burian provided inspiration for six decades and will no doubt continue to do so. Having the opportunity to view his work in Czechoslovakia in autumn of 1992 was an experience chastened by learning that he had passed away some years earlier. To his memory and to those who explored the Gobi to track its prehistoric fauna, Roy Chapman Andrews, Ivan Efremov, and Anatole Rozhdestvensky, this book is respectfully dedicated.

First published 1993
by the author
P O Box 14-421
Panmure, Auckland

ISBN 0-9597983-2-3

Designed by Pages Literary Pursuits
Typeset by Typocrafters Ltd
Scanning by Microdot Media Imaging
Preprint processing by Crystal Colour Graphics Ltd
Printed by Academy Interprint Ltd
Auckland, New Zealand.

This book is printed on acid-free paper.

A Gobi Traveller's Tale

At Earth's Greatest Barrier,
pass beneath the ancient portal
at the Northern gate,
into that forlorn sea of sighs,
where the winds never die,
and demons await.
Follow the Road of the East,
stray not from the path,
'til the tortoise sentinel you meet,
there, at Karakorum,
echo the voice of ten thousand horsemen,
hail the ghost of Genghis Khan.
Now take the Road of the Wind
that runs from Xanadu,
o'er hills of shifting sands,
through glittering valleys of stone,
even past the unfathomable hidden lake,
yet ever on must follow,
'til is found the place you seek,
where seven ages lie beneath your feet,
then look yonder still
beyond the dunes,
through the veil that time has wrought,
there –
where the four winds meet,
and the dragons walk . . .

Contents

Introduction **vi**

Chapter 1. Deserts and Dinosaurs **1**
1.1. A Desert Odyssey *1*
1.2. The Age of Dragons *6*
1.3. The End of the Mesozoic *17*
1.4. Dinosaurs and Birds *20*

Chapter 2. Philosophy, Geology, Palaeontology **27**
2.1. Philosophers and Fossils *27*
2.2. Against the Deluge *29*
2.3. The Ages Beneath Our Feet *31*
2.4. Dinosaurs and Men *35*

Chapter 3. To the Ends of the Earth **39**
3.1. On the Trail of Ancient Man *39*
3.2. The Desert Beyond the Wall *41*
3.3. Where the Dinosaurs Hid Their Eggs *46*
3.4. Mammals from the Dunes *62*

Chapter 4. Road of the Wind **67**
4.1. The Muscovite Campaign *67*
4.2. Armoured Dragons *70*
4.3. Nemegt, Valley of the Dinosaurs *76*
4.4. The Beast of Indrik *92*

Chapter 5. Hunting for Dinosaurs **99**
5.1. The Asiatic Dinosaur Rush *99*
5.2. Exploring the Lost World *101*
5.3. Dinosaur Enigmas *112*
5.4. Desert Quest *123*

Index **136**

Introduction

In 1922, weary members of an American Museum anthropological expedition lead by explorer Roy Chapman Andrews were searching for the origins of early hominids in the uncharted wastes of battle-wracked Mongolia and neighbouring Manchuria. By September the threat of heavy winter snows that would soon render desert routes impassable forced the group back towards Peking. While searching for an old trail in southern Mongolia they unexpectedly stumbled upon what was later proven to be an abandoned dinosaur nesting colony. At the base of a labyrinth of sandstone formations they exposed many unhatched eggs arranged in circular, nest-like patterns, while nearby evidence indicated that the parents had not only been in close proximity, but had apparently nested uninterrupted at the site for countless generations. Thus began one of the most notable chapters in the history of palaeontology.

Disrupted by local political instability and later by World War II when the Gobi was occupied by the Japanese, our story resumes with the discoveries of the 1946 Russian Expedition to the vast Nemegt Valley, an isolated "Lost World" hidden in the Gobi's wild southern sector. During three seasons of field work lead by palaeontologists Ivan Efremov and Anatole Rozhdestvensky, what is possibly the world's most significant repository of late Cretaceous dinosaurs was explored. Some 120 tonnes of skeletal material was removed, much of it from the sensational "Dragon's Tomb" cemetery where imprints of remarkably recent-looking dinosaur skin were discovered in dried mud.

Subsequent decades saw the search continue at Nemegt and other locations with a series of particularly successful joint Polish-Mongolian expeditions under the leadership of Zofia Kielan-Jaworowska, as well as later Russian-Mongolian field work involving palaeontologists like Sergei Kurzanov and Rinchen Barsbold. Among the many new finds were the unusual sloth-like deinocheirids, dome-headed pachycephalosaurs, and evidence of giant sauropods which had survived extinctions that claimed similar dinosaurs elsewhere.

Now, 100 years after the first Mongolian fossils were described by the geologist Vladimir Obruchev, the story of the search for dinosaurs in the Gobi Desert is presented in Chapters three, four and five of *Dragons from the Dunes*. Along with details and narratives of individual expeditions are over 200 figures, illustrations, maps and photographs, many published for the first time. The evidence that not all dinosaurs died at the close of the Cretaceous is outlined in the first of two introductory chapters that also review dinosaur classification and the history of palaeontology prior to the events later recounted.

Though the harsh terrain of Asian deserts had always dictated that travellers and merchants keep to known caravan routes, these evocative, sweeping landscapes captivated the restless temperaments of many explorers, notably Marco Polo and Nikolai Przhevalski. Not content to permit the sprawling *terra incognita* scripts of early maps to go unchallenged, they were always inspired by the promise of discovery, be it past the next stone-strewn valley or beyond a hazy mountain range upon a distant azure horizon. Join us to undertake a journey through a timeless desert, in the words of Roy Chapman Andrews "to the ends of the Earth" in search of the Gobi's most elusive inhabitants, Mongolia's Dragons from the Dunes.

Chapter One

Deserts and Dinosaurs

1.1. A Desert Odyssey

"WALL of Spears" – "Desert of Sand" – "Great". These are said to be just some of the various translations in different languages of the word "Gobi". Although the original meaning appears to be lost in antiquity, all peoples of the Asian continent use the word in reference to the vast arc-shaped plateau that straddles the high eastern Central Asian tableland, between the elevations of 1600 and 2000 metres. Now known to have been the subtropical domain for a great variety of dinosaurs during the Mesozoic Era, the Gobi today constitutes a desolate area of arid, stony desert and semi-arid rolling grassland steppe, one of the Earth's last remaining truly wild places.

Some Western geographers allocate a length of 1600 kilometres to the Gobi (east to west), with a total area of some 1.3 million square kilometres. Such measurements assume its northern reaches to be bound by the Altai and Hangayn Nuruu mountains, and its western ones by the eastern peaks of the Tien Shan ranges. In fact, the desert's western extent is limited only by local use of the word "Gobi", and other writers have given the length as extending from the Pamir mountains south of Kazakhstan through to the inner confines of Manchuria in the east, a distance of more than 5000 kilometres. Whatever its true boundaries, a huge portion of the Gobi falls within the borders of Mongolia, which at 1,565,000 square kilometres, is one of the world's largest land-locked countries that is geographically wedged between Russia to the north and China to the south.

The inhabitants of Mongolia use the word "Gobi" to describe a particular order of land-form features characteristic of the desert; shallow and wide topographic basins whose floors are most frequently composed of gravel, less frequently of sand or pebbles. Often described as the wildest and most infertile section of Asia, the Gobi today has an average density of just one person per square kilometre. Around some of the better- watered northern basins of the Selenga and Orhon rivers,

Although described as "desert", the Gobi varies geographically from very arid sandstone basins (top) *to semi-arid rolling grassland steppe* (bottom). *(Top by Z. Kielan-Jaworowska, courtesy of W. Skarzynski, Polish Palaeobiological Institute, bottom courtesy of Adventure World, Auckland)*

Above all else, nomadic Mongols value their animals and the freedom to roam the steppes. (Courtesy of Adventure Travel Shop, Auckland)

the density may approach four times that average, but a great many of the harsher sectors are almost entirely devoid of inhabitants. With one fifth of Mongolia's approximately 2 million people living in the capital Ulan Bator, most of the remaining population are desert-dwelling shepherds whose characteristic abode is the circular, felt-lined "yurt" tent. These people are the descendants of the well-drilled horsemen of the Mongol leader Genghis Khan (1162-1227), whose mighty Tartar Empire of conquered peoples and amassed lands stretched across a third of the earth's circumference, from the Danube in Hungary to the Yellow River in China. Today, most Mongols still lead a nomadic life with small family groups continually on the move in search of scanty pastures for their animals, dismantling and carting their yurts with them. Livestock outnumbers humans by thirteen to one in Mongolia (Mongols are the world's highest consumers of meat) and in addition to sheep and cattle, other domestic animals tended include goats, long-haired yaks, Bactrian camels, and Mongolian horses, while well-trained dogs warn of approaching wayfarers.

The wildlife of the desert is sparsely distributed but diverse, indeed some species are to be found nowhere else. The open steppes are the habitat of the marmot, the leaping jerboa and other small rodents, as well as the legendary Mongolian dzheiran gazelles which are the world's fastest animals over long distances. They have been clocked at seventy kilometres per hour and unlike that other swift-footed mammal, the cheetah, are able to maintain their stamina for a considerable time. In semi-arid areas, never very far from water-holes can be found mountain sheep and large-horned ibexes, wild horses ("takhs"), Asiatic wild asses ("kulans"), Bactrian camels ("khavtgays") and perhaps even a wolf or Gobi bear ("mazalai"). As in other deserts, the hot, dry climate of the Gobi is the habitat for many of the smaller species of lizards and snakes, particularly vipers, while the Gobi scorpion (*Buthus martensi*) and biting-spiders such as the giant solpug genus *Galeodes* add to the list of species to be treated with respect.

To the Easterner, the Gobi has long been considered a place of exile and banishment, and at the western-most portal of the Great Chinese Wall can be found the archway of the "Gate of Sighs" which looks out onto the desert to the northwest. The walls here are covered in lamentations and verses that have been etched into the stonework by exiles, disgraced officials, and common criminals who were no longer to be tolerated within China's borders. To Westerners, the Gobi appeared to be a bewildering and mysterious place that revealed few of its secrets and was very capable of exacting a harsh toll on the unwary or ill-prepared traveller. Early explorers returning to the west spoke of violent and terrifying sand storms or "burans", and of vast fields of ceaselessly drifting, treacherous sands described by the local inhabitants.

The great explorer Marco Polo (1254-1324) was the first non-Easterner to describe the Gobi in any detail. Between 1275 and 1292, he travelled extensively throughout the Tartar Empire during the reign of Kublai Khan (1216-1294). This was the period when the empire established by Genghis (the grandfather of Kublai), had reached the peak of its organisation. Returning to Venice in 1295 with many stories of the splendours of Central Asia, Marco Polo later recorded of the Gobi "....the length of the desert is so great that it would take a year and more to ride from one end of it to the other. It is all composed of hills and valleys of sand". He went on to speak of the many spirits that haunted the desert wastes, calling men's names out aloud and giving rise to strange noises that to the weary traveller, can sound deceptively like the clang of drums or the clatter of horses' hooves.

A desert phenomenon that fosters many demonic beliefs is the dust-spout, small spirals of wind-driven sand that sometimes appear over flat ground, often in large numbers. Called "kwei" or "dust-demons" by superstitious Gobi-dwellers, the pillars of sand give the vague impression of an unseen being "walking" as it continually folds a garment of sand about itself (the body form is thought to be male or female depending on the direction of rotation). Inhabitants of the Lob area believe the kwei that haunt sand-hillocks can take the form of a black eagle that is responsible for producing the desert voices or "azghun" so often blamed for luring wayfarers away from caravan routes. To those inexperienced in Gobi travel, straying from the road even a short distance in poor light can prove fatal, and many have perished not only by heeding the sounds themselves, but also by going in search of companions who have.

During the 19th century, several systematic journeys through sectors of the desert were undertaken by notable explorers including Pavlinoff and Matusovski (1870), Ney Elias (1872-1873), von Richthofen (1873), and Nikolai Przhevalski, the Russian colonel who made extensive traverses of the southern sectors (1870-1877), and is celebrated for his discovery there of the world's last remaining true breed of wild horse which now bears his name, *Equus przewalskii*. The first accurate seasonal temperature readings taken in the Gobi surprised European geographers. Because of the acutely continental climatic conditions that it experiences, the Gobi is subjected to greater extremes of temperature variation than any other desert. Summers are intensely hot, with July temperatures reaching 45°C, while in January, during the dead of winter when the plateau is swept by bitter snow-bearing winds from the Arctic Ocean, temperatures plummet harshly to -40°C.

Yet as timeless as the desert landscape may superficially appear, it is believed that the arid and semi-arid conditions now found on the

Gobian plateau are the result of relatively recent climatic and geographical changes, developing perhaps as late as some 50,000 years ago. But the times of the Mesozoic Era (Age of Middle Life), with which we are concerned are of course far older, encompassing the time span from 65 million years to 225 million years ago. During the last 135 million years, the Central Asian region has never been completely inundated by the sea. But during the latter part of the Mesozoic when the dinosaurs reached their greatest diversity, the sea did encroach deep into eastern Central Asia and areas that are now situated far inland then lay closer to the coast. Then, the Gobi was no desert, it was a land of lakes and rivers, of forests and bush-clad rolling steppes and floodplains. Lying at a relatively low elevation, it was well-watered due to drainage from upland areas to the west. The topography resulted in a large network of meandering and slowly running rivers that flooded large areas, producing swamps, marshland, and shallow but expansive lakes up to ten metres in depth. Fossil and other evidence from late Mesozoic strata reveal that the area enjoyed a coastal, subtropical and humid climatic regime.

This combination of equitable geographical and climatic conditions with an abundant supply of vegetation permitted the Gobi to sustain large numbers of a wide range of dinosaurs, as well as other lesser reptiles such as crocodilians and turtles. The sedimentary deposits that collected at the bottoms of lakes and swamps as well as sands and clays transported along rivers have preserved many skeletons from this rich reptilian fauna of the late Cretaceous. These fossiliferous deposits

have since hardened into sandstones and shales, and due to the erosive forces of wind and rain, often outcrop along the banks of ravines, gorges, and cliff faces, making them reasonably accessible to fossil-hunters.

Since the 1920s, palaeontologists and geologists from four countries – the United States, Russia, Poland and Mongolia – have undertaken extensive expeditions into the harsh interior of the Gobi Desert. Often working under difficult and sometimes life-threatening conditions, these expeditions have excavated remarkable collections of Cretaceous dinosaur remains, as well as other reptile and mammal fossils from both Mesozoic and more recent strata. These finds, many of which have represented completely new taxonomic groups, significantly broaden our knowledge of the classification, behaviour and ecology of dinosaurs and other extinct forms. A further consequence of such field-work is that dinosaurs are today enjoying something of a renaissance with both the public and academic alike. Every year, an increasing number of people visit internationally renowned museums to view dinosaur remains recovered from areas as diverse as the lowlands of Europe, the great western plains of the United States, the coastal forests of East Africa and the deserts of Central Asia.

In the past, the dinosaurs best known to the public have been those found during the late 19th century in the rich North American Mesozoic strata during the pioneering era of dinosaur palaeontology. These classical dinosaurs have been reproduced in countless texts, popular books and animated fiction films, with the result that few children cannot readily recite from memory names like *Stegosaurus, Allosaurus* or *Diplodocus*. To a large extent, such dinosaurs have traditionally been viewed with an undue bias. Treated as cumbersome, primeval monsters far removed from more familiar animals, they were often portrayed as overgrown reptiles that crashed their way through tropical forests when driven by hunger, or spent their mundane existence wallowing in swamps to relieve the otherwise unbearable weight from their limbs. The largest species represented the extremes of evolution's wild experiments, and could only drag their huge bodies ashore to lay their eggs before being forced to return to the relief of the swamps. Regarded with a mixture of awe and curiosity, dinosaurs were thought to have become extinct due to their large size, slow and clumsy habits, and inability to change, being apparently doomed to failure by their very design. In the United States, the word "dinosaur" is often used synonymously for products the public considers as too large or outdated.

We now realise that this concept of dinosaur life is completely inaccurate and is itself obsolete. Dinosaurs were among the most successful vertebrates ever, reigning for an unparalleled span of some 165 million years. Obviously highly adaptable, they produced a great diversification and evolved at the expense of the advanced mammal-like reptiles, as well as the early mammals which only began to diversify after the dinosaurs' extinction. Although many dinosaurs did indeed achieve monstrous proportions, the average size of all dinosaurs combined is not far from that of a human. Even the sauropods, which were the largest types, found their size to be no hindrance and are represented in the fossil record over a period of no less than 125 million years. Their limb construction, like those of other dinosaurs, reveals a mammalian-like parasagittal efficiency of movement more akin to the grace of a giraffe than the scuttle of a crocodile. Some dinosaurs are known to have had communal nesting and parental habits that hint at a more complex social behaviour than is characteristic of most modern reptiles. Indeed in many respects, dinosaurs appear to have acted more like mammals or birds than like typical reptiles. Considered in this light, questions surrounding the causes of their extinction are only relevant when considered against the reasons for their success over such a long period of time.

But in this technological age when Man has reached out to the other planets of the solar system and is capable of living outside the Earth's environment in orbiting space stations, some may question the value of studying the remains of animals that had already died out 65 million years before the appearance of *Homo sapiens*. Naturally, palaeontology attempts to explain how extinct animals like dinosaurs evolved, lived, interacted, reproduced, and died out. But it is not only a study of animals in their own right; more than ever before it relates directly to the course that Man's future may take, and that of the other species with which he shares this planet. In her book *Hunting for Dinosaurs* (1969), Prof. Zofia Kielan-Jaworowska, leader of the Polish-Mongolian Expeditions to the Gobi in the 1960s and early 70s writes...

"The study of animals that lived on Earth millions of years ago is not merely a study of their anatomy, but first and foremost a study of the course of evolution on Earth and of the laws that govern it. All the philosophical systems ever developed by Man are essentially anthropocentric, with the tragic consequences for humanity which we know from history and which we are experiencing today. Studying the evolution of animals that inhabited Earth for millions of years and comparing this with the history of Mankind – so short on the geological scale – put Man's position in the living world in its proper perspective and help counteract anthropocentric ideas...

"If the history of an animal group is known, it is possible to tell which species are currently dying out, and will die out completely unless put under special protection. Extinction of species is just as common as their creation – these are merely two sides of the same process. Since the extinction of species and of large animal groups is a common evolutionary process, it is natural to inquire whether it might not in future affect those groups that are now at the peak of their development. Study of the reasons behind the extinction of species on our planet and of the laws governing this process may yet prove to be highly important to our own species in shaping its future development on Earth."

The remainder of this chapter as well as chapter two provides a summary of the rise of the dinosaurs, their classification, environmental conditions during the Age of Reptiles, the relationship between dinosaurs and birds, and finally, the development of the science of palaeontology and how it is used to decipher the fossil record. Chapters three, four and five contain chronological narratives of the preparation, field-work and results of the various expeditions that have been undertaken to the Gobi Desert.

1.2. The Age of Dragons

The Mesozoic has been divided into three periods, the Triassic of 25 million years duration, the Jurassic of 65 million years, and the Cretaceous of 70 million years. This was an important time for reptilian development and evolution as they occupied and dominated many available habitats throughout the world. Consider the spectrum of surviving reptiles, ranging from five-centimetre-long wall lizards that can be held on a finger, to nine-metre-long estuarine crocodiles and thirteen-metre-long anacondas capable of devouring an entire deer.

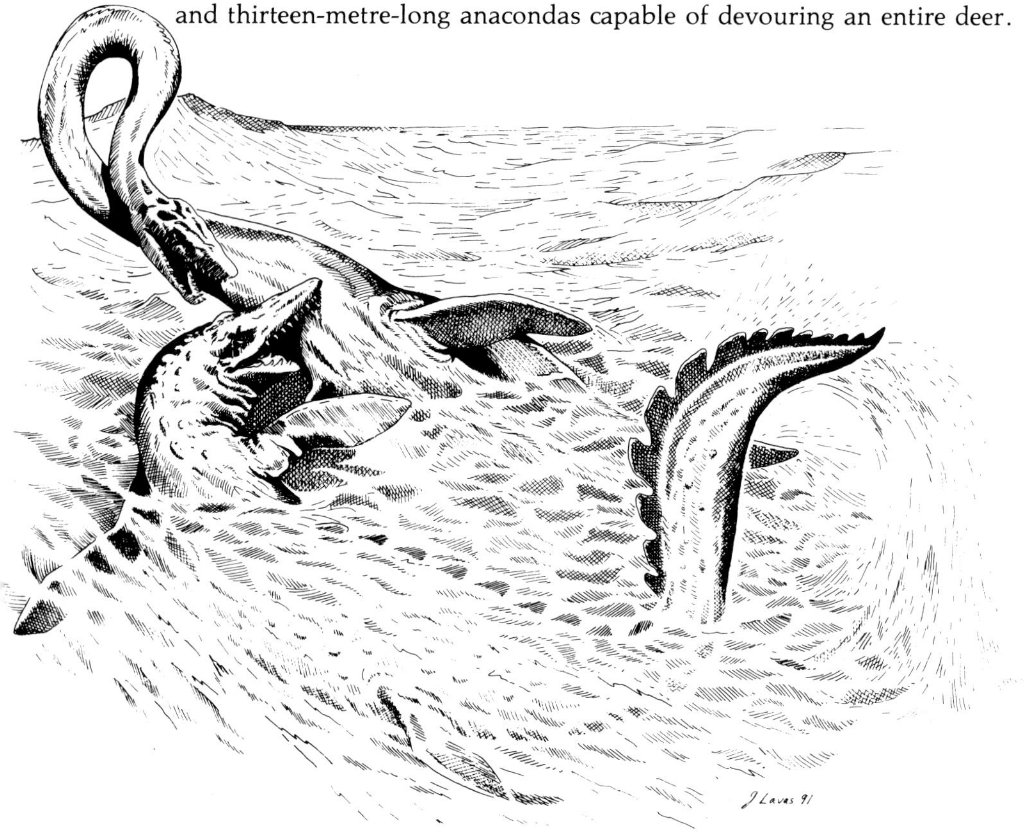

During the late Cretaceous, the inland sea of Kansas was the scene of ferocious conflicts between the long-necked plesiosaur Elasmosaurus *which reached lengths of thirteen metres, and the crocodile-like mosasaur* Tylosaurus, *a voracious predator that used its laterally-compressed tail to propel its eleven-metre body through the water. Skeletons of these giant marine reptiles often bear the scars and healed fractures of such encounters. (Based on Z. Burian)*

These would represent only a fraction of the range of reptiles that were present during the Mesozoic.

Even before the Mesozoic, mammal-like reptiles had become the dominant land animals during the Permian period of the late Palaeozoic Era (Age of Ancient Life). The primary reason for the success of the reptiles was due to their development of the amniotic egg, a feature also possessed by birds and early mammals who themselves later evolved from the reptiles. The first reptiles date from the Carboniferous period, and were the descendants of amphibians. Like today's frogs, newts and salamanders, Carboniferous amphibians were always tied to the aquatic environment because of their method of reproduction. The amphibian egg was and is unprotected by a hard shell, and it dries out quickly if removed from water. The amphibian embryo takes in oxygen and nourishment from the surrounding water, and excretes its waste back into it. Amphibians need to lay large numbers of eggs to ensure that a small number will survive, because not only are the tiny, soft eggs at the mercy of pond predators, but after hatching, the undeveloped tadpoles are easy prey for insects, fish and even other amphibians.

At some stage during the Carboniferous, some amphibians began to lay hard-shelled eggs that provided for all of the embryo's requirements. Food was supplied by a relatively large yolk sac attached to the digestive tract while the embryo itself floated within the fluid-filled amniotic membrane. This membrane enclosed the young animal in its own private pond and acted as a buffer against injury and desiccation. From the rear of the embryo's body grew a small tube connected to the allantoic sac in which the body's waste was collected. The embryo, the yolk sac, and the allantois were all further surrounded by another membrane, the chorion, and finally, enclosing all of this was the smooth, protective shell. Although firm in construction, the shell was nonetheless porous and allowed the exchange of gases via blood vessels connected to the allantois. This system functioned as an efficient lung to the outside world, taking in oxygen and excreting carbon dioxide. No longer was there a need to lay many unprotected eggs, as a small number of amniotic eggs well-hidden within undergrowth at the edge of a pond would be reasonably safe from predators. Animals that laid such eggs would today be classified as reptiles. When the climate eventually changed and swamps dried up, it was these animals which were best suited to survive.

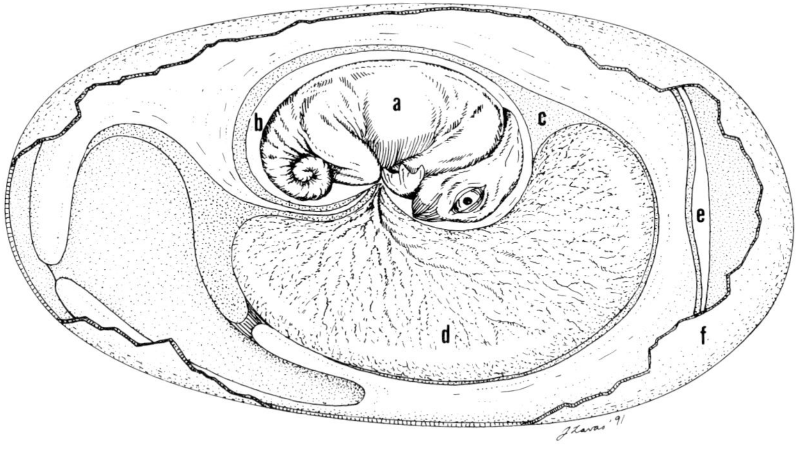

Hypothetical reconstruction of an embryonic protoceratopian dinosaur developing within the protection of its fluid-filled amniotic egg. a=embryo, b=amnion, c=allantois, d=yolk, e=chorion, f=calcareous shell. (Based partly on G. Caselli)

This is not to say that the reptiles remained land-bound, for many later returned to the water. By the Jurassic and Cretaceous periods, the world's extensive and largely non-seasonal seas swarmed with a myriad of unusual reptilian forms. These included the reptilian analogues of modern dolphins — the ichthyosaurs, long and short-necked plesiosaurs of all sizes, (which have been described as having the body of a turtle with a snake threaded through it), and giant, crocodile-like marine lizards of the Mosasauria group which were related to today's monitor lizards of the Old World.

But on the land, and in lakes, rivers and swamps, as well as in the Mesozoic skies, it was reptiles of the subclass Archosauria that were dominant, both in terms of numbers and adaptation. Five orders comprise the Archosauria, or "ruling reptiles", four of which are now extinct. These are 1) the Thecodontia or "socket-toothed" reptiles, 2) the Pterosauria or flying reptiles, 3) the Saurischia or "reptile-hipped" dinosaurs and 4) the Ornithischia, or "bird-hipped" dinosaurs. The fifth order that still survives today, though in a much depleted state, is the Crocodilia, which comprises the crocodiles, alligators, caimans, and gavials. In addition, due to similarities between the morphology of birds and that of thecodonts and saurischian dinosaurs, the class Aves is often included within discussions of the archosaurs. Archosaurs rose to prominence in the Triassic during the early stages of the Mesozoic, and maintained their dominant positions until the end of the era. The evolutionary radiation that they underwent was exceptional, and finds comparisons with the earlier evolution of the mammal-like reptiles as well as the evolution of the mammals that later took place during the Cenozoic Era (Age of Recent Life) after most of the archosaurs had become extinct.

The oldest known reptilian ancestors of the archosaurs were previously believed to be the cotylosaurs which date from 330 million years ago during the Carboniferous. It has recently been shown however, that this group actually included some forms that were amphibians. The oldest known group of undoubted reptiles (previously classified as cotylosaurs) are now classified as captorhinomorphs. Ungainly and bulky omnivores characterised by a sprawling gait and limbs positioned laterally out from the body, they are believed to have given rise to the Eosuchia, and these in turn to the earliest archosaurs.

In addition, these "stem" reptiles also constituted the stock from which evolved the first of the mammal-like reptiles, the pelycosaurs. Forms such as the carnivorous *Dimetrodon* and the herbivorous *Edaphosaurus* were typical pelycosaurs with long spine-supported "sails" along their backs that are thought to have acted as solar heat

The herbivorous pelycosaur Edaphosaurus *from the late Carboniferous and early Permian. (Based on Z. Burian)*

regulators. Pelycosaurs still retained a relatively poikilothermic metabolism, that is, like many modern reptiles they depended to a large extent on externally derived heat to maintain their body temperature. It was not until the latter half of the Permian, over 225 million years ago, that homoiothermic mammal-like reptiles evolved.

These were the therapsids, and they maintained a constant body temperature by internal metabolic processes. Their remains are well known from Permian and Triassic formations particularly in Russia, China, South-Africa, Laos, the West Indies and Antarctica. Therapsids included both herbivorous forms, such as the lumbering *Kannemeyeria* and the hippo-like *Lystrosaurus,* as well as fast, agile carnivores like the dog-sized *Cynognathus* and the sabre-toothed *Lycaenops,* a highly active, wolf-like carnivore whose gracefully constructed skull and limbs were very similar to those of mammals. Even though therapsids were most probably oviparous egg-layers like their captorhinomorph ancestors, they were an advanced and highly successful group of animals that maintained dominance throughout the late Permian and on into the Triassic. By this time, they had begun to die out following the rise of the archosaurs. They did not become extinct without leaving descendants however, and the first "true" mammals — hair-insulated, homoiothermic and able to suckle their young, arose toward the end of the Triassic. Throughout the next 115 million years or so, mammalian development was very slow and extremely modest however, resulting in only several orders, most members of which only managed to attain the size of a rat, or at most a cat. It is very likely

Lycaenops, *one of many highly successful mammal-like therapsid reptiles from the late Permian. (Based on Z. Burian)*

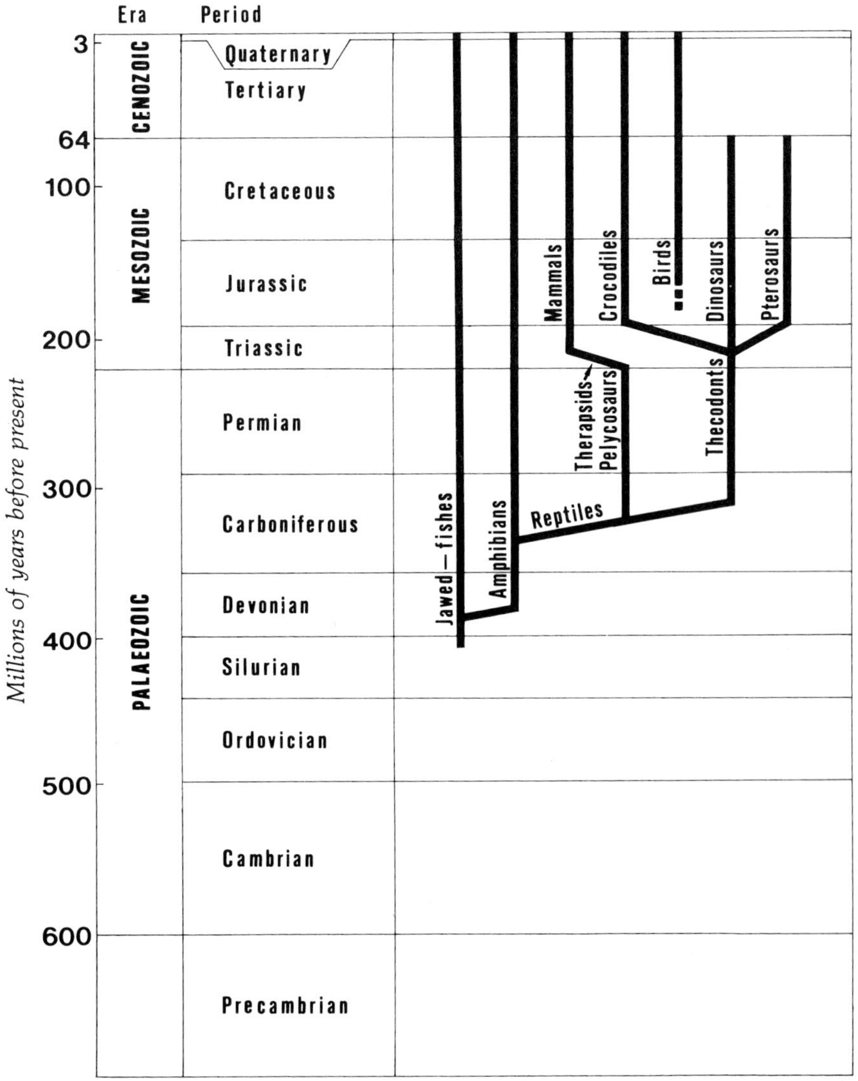

Geological time chart with schematic representation of the fossil histories and assumed relationships between the various vertebrates (note that although represented by one bar, the "dinosaurs" contain two distinct orders).

that these early mammals lost the initial competition to the rapidly rising archosaurs, and consequently were destined to play only minor, nocturnal roles in the ecology of the Mesozoic environments.

During the Triassic, the descendants of the captorhinomorphs also gave rise to the earliest archosaurs, the thecodonts. The name "thecodont" refers to the teeth being fixed in sockets, a shared feature among archosaurs as well as some other reptile groups. Thecodonts have been divided into the proterosuchians, the pseudosuchians, the aetosaurs, and the phytosaurs. Representatives from these groups varied greatly from light and fragile gliding forms such as *Longisquama* (a pseudosuchian), to armoured forms such as *Desmatosuchus*, an aetosaur which possessed long spikes out the sides of the neck, and to the crocodile-like phytosaurs that in spite of their appearance were not ancestral to the crocodilians. Rather, it appears that the rise of the crocodilians may have led to the decline of the phytosaurs which were the chief freshwater predators of the Triassic. Phytosaurs are easily recognizable from early crocodiles due to their nostrils opening just in front of the eyes, rather than at the end of the snout. Other thecodonts such as the pseudosuchian *Saltoposuchus* from the late Triassic, were small, lizard-like reptiles with narrow skulls, strong hind legs and long tails which they used to balance themselves during bipedal locomotion. Fast and agile predators, these thecodonts exhibited a morphology that was to be increasingly adopted by later archosaurs. The main developments were to be found in the limb and girdle constructions which led to major improvements in locomotor abilities, a factor which is thought to have played no small part in the archosaurs' early success in the Triassic.

While other reptiles simply retained or improved upon the quadrupedal locomotion that was characteristic of their ancestors, the archosaurs tended strongly toward bipedalism and a semi-erect gait. This was accompanied by the development of a more powerful caudal structure, with the long tail acting as a balance to counteract the mass of the body which tilted forward at some degree to the pelvis. The sacrum was also strengthened, and extra ribs were developed in this area. The alterations that took place in the limbs were very significant. Whereas the ancestral condition involved the femur and humerus being positioned in a splayed out fashion, the archosaur femur tended inwards to a nearly ventral position. This enabled the hindlimbs to bear the full weight of the body, thus permitting a powerful, parasagittal and mammalian-like limb movement. Another consequence of the bipedal position was a lessening of the locomotory importance of the forelimbs, these often tending to degenerate, even leading to the loss of digits as in the case of some later carnivorous dinosaurs. A complete reliance on carnivorous diets was also lost, and the simple, pointed teeth of early archosaurs were, in some cases, eventually replaced by teeth modified for herbivorous diets, or even by a horny, edentulous "beak" as in the horned dinosaurs.

The environments in which the archosaurs and their contemporaries lived greatly influenced their evolution. Up to the beginning of the Triassic, all the world's continental landmasses are believed to have still been united in a single, giant supercontinent named "Pangaea". Along its eastern sector, a great tropical ocean named the "Tethys" (in Greek mythology Tethys was the wife of Oceanus and mother of the seas) encroached deeply into the middle of Pangaea separating it into a northern portion named "Laurasia" and a southern portion named "Gondwanaland". Laurasia comprised lands that today make up the continental areas of North America, most of Europe, and Asia, while to the south, Gondwanaland comprised South America, Africa, India, Antarctica, and Australia, as well as sections of the Middle East and Western Europe. At the beginning of the Permian, an extensive ice-cap had covered parts of the southern landmasses, and in the accompanying cooler climates, poikilothermic reptiles were at a disadvantage compared to the more homoiothermic mammal-like reptiles.

But by the Triassic, the climate had stabilised and subtropical conditions prevailed over virtually all of Pangaea. There were no ice-caps at either pole and forest growth extended to latitudes where today no vegetation grows. Reptilian life abounded, especially in the warm Tethys Ocean into which reptiles had not previously ventured. Many groups of freshwater fishes had returned to the seas by this time, and they in turn were followed by reptiles which fed upon them. The dolphin-like ichthyosaurs developed bodies perfectly suited to fast ocean swimming, and whereas they had previously laid eggs on land, they now gave birth to live young in the water. Other reptilian groups that took advantage of the equitable environments of the Tethys included the nothosaurs, semi-aquatic animals with webbed feet and finned backs, and the placodonts, heavily-built marine reptiles with armoured abdomens and teeth suitable for crushing shellfish. The

The late Jurassic ichthyosaur Eurhinosaurus, *one of many so-called "fish-lizards" that cruised the warm Mesozoic seas. (Based on Z. Burian)*

Jurassic period was a time of continued climatic stability with the Earth becoming wetter and warmer. The flora became truly tropical, rich in conifers, cycads, horsetails and ferns, while the ginkgo tree (of the same genus as the modern maidenhair tree) became characteristic of Jurassic landscapes. In the humid and calm skies, the reptiles that were to prove supreme were the pterosaurs.

Pterosaurs are divided into rhamphorhynchoid (= prow-beaked) and pterodactyloid (= winged-fingered) forms. They were the first reptiles capable of powered flight and arose toward the end of the Triassic, reaching their greatest diversity in the Jurassic. A typical rhamphorhynchoid pterosaur was the genus *Rhamphorhynchus* known from Jurassic rocks in Germany, England, Portugal and East Africa. These were small hair-covered reptiles, with wing-spans of up to 1.75 metres, and were equipped with long tails terminating in a rudder-like stabiliser. Fast and agile fliers, the rhamphorhynchoids lived around lakes and lagoons where they utilised their forward-projecting teeth for capturing small fish. The best known of the pterodactyloid pterosaurs is the genus *Pteranodon* from the late Cretaceous of North America and around Russia. With a wing-span of almost seven metres, this finely-boned and toothless reptile cruised coastal areas such as the inland sea that covered the state of Kansas, feeding on fish and crustaceans, and possibly laying its eggs on selected "nesting" islands. A more recently discovered genus is *Quetzalcoatlus* from the state of Texas. This long-necked type had an even larger wing-span, 11.5 metres, and is thought to have had vulture-like habits, feeding on dinosaur carcasses on the open plains.

The late Cretaceous pterosaur Pteranodon. *(Based on Z. Burian)*

Although the Jurassic was a time of general stability, it was also the period during which the landmasses of Pangaea began to separate into individual continental plates that gradually rifted apart. Many groups of reptiles that had previously flourished now began to decline, including the marine nothosaurs and placodonts, the freshwater-dwelling phytosaurs, and the terrestrial mollusc-eating rhynchosaurs. Ichthyosaurs and long and short-necked plesiosaurs now shared the warm Jurassic oceans with the teleost (true-boned) fishes, as well as sharks and rays. The landscape of the Jurassic was typically flat, poorly drained, and divided by sluggish, meandering rivers and by expansive areas of lakes, swamps and marshland. Shallow but extensive tropical lagoons spread all the way from Portugal to France and southern Germany in Europe, while large sections of North America and eastern Africa formed warm and humid swampy floodplains. It was in such environments that the most successful of the archosaurs, the two orders of dinosaurs, were to rapidly diversify.

Saurischian dinosaurs

The saurischian dinosaurs were the first to appear during the Triassic, and they were small and active bipedal carnivores called coelurosaurs (= hollow-tailed reptiles). These agile dinosaurs varied from one to three metres in length and had long, flexible necks and small heads with pointed snouts. Well known examples included *Compsognathus*, *Coelophysis*, and *Ornitholestes*, while their remains have been found variously within Triassic, Jurassic and Cretaceous deposits of North America, Africa, Australia, Asia and Europe.

Coelophysis. *(Based on L. Darling)*

The oviraptorosaurs (= egg-thieving reptiles) – also sometimes known as oviraptors – were an unusual and somewhat poorly known group containing the late Cretaceous Gobian genus *Oviraptor*. The caenagnathid dinosaurs, known from fragmentary North American remains, should probably also be included with the oviraptorosaurs. Like the coelurosaurs, the body shape of oviraptorosaurs was of slender build, with a length of between 1.5 and two metres, while the skull was short-snouted and ended in a toothless parrot-like beak. More material from these bipedal dinosaurs have recently been described by the Mongolian palaeontologist Rinchen Barsbold.

From Gobian and western North American Upper Cretaceous strata, the two-metre-long troodontids, originally known as the saurornithoidids (=bird-like reptiles), formed another group of slender theropods. They were distinctive due to their extraordinarily large brain capacity when compared to that of other dinosaurs. This was most likely a result of highly developed reflex and reaction capabilities which when combined with their unusually large eyes, suggest that these bipedal reptiles probably hunted nocturnally. Perhaps even the small Mesozoic mammals were their prey. Previously named genera included *Troodon*, *Saurornithoides*, *Stenonychosaurus*, and *Tochisaurus*, though some of these may be synonymous.

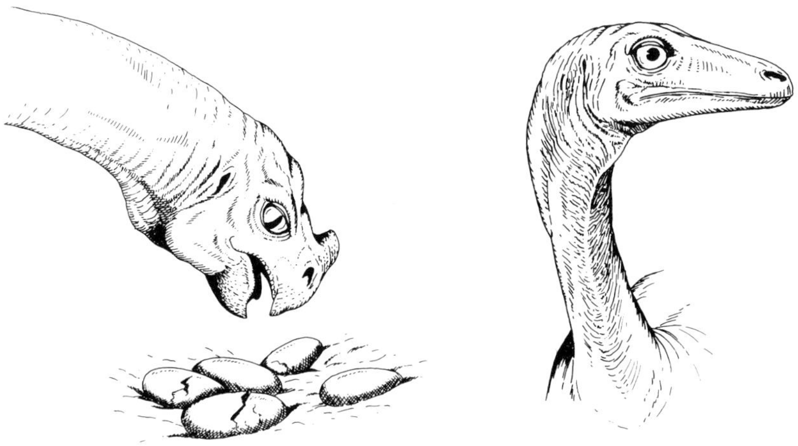

Left: Oviraptor. Right: Troodon. *(Based on J. Sibbick)*

The fourth and last group of small theropods, the dromaeosaurs (= swift-running reptiles) are, like the oviraptorosaurs, rather poorly known. Measuring 1.8-3.3 metres long, dromaeosaurs had relatively large heads with powerful jaws and were fleet-footed. The second toe was modified into a large, sickle-shaped claw that may have been used in attacking other dinosaurs. Well known genera included *Velociraptor* (which now includes the former genus *Deinonychus*) and *Dromaeosaurus*, and remains have been found in Cretaceous strata of the Gobi, North America and Europe. Several possible new dromaeosaur genera are known from Mongolia and China.

The medium-sized theropods were represented by the ornithomimosaurs (= bird-mimicking reptiles) or "ostrich dinosaurs"

Velociraptor. *(Based partly on R. Bakker)*

from the late Cretaceous of the Gobi and western North America. These were three to six metres or more in length, and as their name suggests were remarkably ostrich-like in anatomy, even in so far as possessing toothless jaws (except for a recently found Mongolian genus that did have small teeth), long flexible necks and finely-constructed skulls with large orbits. Their long, slender legs were well-suited to fast running over open steppe areas, while the diet was probably

Struthiomimus. *(Based on J. Sibbick)*

omnivorous. Genera included *Ornithomimus, Struthiomimus* and *Gallimimus.* Most authors also assign the deinocheirids, known only from forelimb remains collected in the Nemegt Valley of Mongolia, to this group.

The Carnosauria (= flesh reptiles) were giant-sized predatory theropods varying in length from six to thirteen or more metres. These included the most formidable terrestrial predators ever to evolve. Common in both Jurassic and Cretaceous strata, their remains have been found in Asia (particularly the Gobi), western North America, South America, Europe and Africa. All were heavily-built reptiles with heavy jaws, powerful, pillar-like hind legs and small to vestigial forelimbs. Because many genera are represented only by fragmentary remains, it is not clear how closely the various carnosaurs were related to one another. Examples of carnosaur genera included *Allosaurus, Ceratosaurus, Megalosaurus, Spinosaurus, Tyrannosaurus* and *Tarbosaurus.*

Ceratosaurus. *(Based partly on J. Sibbick)*

A number of late Cretaceous theropods, particularly several from the Gobi, were so unlike other known types that they cannot be confidently classified, and are thus placed temporarily together as miscellaneous theropods. Gobian genera here included are the two segnosaur genera *Segnosaurus* and *Erlikosaurus*, and also *Therizinosaurus*, *Avimimus*, and *Elmisaurus*. The segnosaurs (= slow reptiles) are incompletely known but were peculiar due to the ornithischian-like pelvis. The jaws were toothless and the diet may have included freshwater fish. At this stage, it appears that they will have to be classified in their own infraorder. The therizinosaurs (= scythe reptiles) are known only from fragmentary remains of massive forelimbs and claws, the largest known of any carnivorous dinosaur. Their life habits can only be surmised until more complete remains are found. *Avimimus* (= bird-mimic) is thought to possibly have been a feathered theropod due to its remarkably bird-like skeleton and its forelimb morphology. The remains, however, are too fragmentary for a conclusion to be drawn one way or the other. *Elmisaurus* (= foot-reptile), described by Osmolska in 1981, is a poorly preserved theropod that appears to have had the grasping hands of a dromaeosaur and the fleet-footedness of an ornithomimosaur. *Elmisaurus* will probably constitute a new family.

The last major group of saurischian dinosaurs to be considered is the Sauropodomorpha (= foot reptiles) which included the prosauropods and the sauropods proper. Known from all continental areas except Antarctica, the prosauropods were abundant from the late Triassic to the early Jurassic, and were medium to large dinosaurs from 2.5 to eight metres in length. Able to walk quadrupedally or to stand bipedally, the prosauropods had heavy bodies, long necks, small heads and hindlimbs much stronger than forelimbs. This was the first major dinosaur group to exploit plant food (most archosaurs of the Permian and Triassic having been carnivores), and they were the evolutionary forerunners of the larger sauropods. The genera *Plateosaurus*, *Anchisaurus* and *Yunnanosaurus* were representative prosauropods.

The largest and heaviest of dinosaurs were contained within the Sauropoda. Also found in most continental areas (except Antarctica), they reached their peak of development in the Jurassic, though late Cretaceous genera such as *Quaesitosaurus*, *Opisthocoelicaudia* and *Nemegtosaurus* from the Gobi show that the sauropods survived until the end of the Mesozoic. Typically "brontosaurus" like in appearance, sauropods had long, whip-like tails and necks, diminutive heads, and column-like legs supporting very deep and heavy bodies. Some, like *Brachiosaurus*, weighed over eighty tonnes in life, while others were more slenderly-built, such as *Diplodocus*, which reached lengths of over twenty-three metres. Another slender sauropod was *Mamenchisaurus* from China, of which the neck alone accounted for ten metres of length. Whether sauropods were partly aquatic or fully terrestrial in habits is often debated. It has been suggested that different types of sauropods probably showed a spectrum of life habits ranging from terrestrial to semi-aquatic. Sauropods would certainly not have been able to withstand the enormous pressures on the lungs were their bodies completely submerged as has often been depicted in older reconstructions. A more likely position for species that may have been semi-aquatic would be floating in shallow waters with the lungs close to the water line, and there are several fossilised track-ways which show that sauropods had kicked their way along whilst floating in this manner.

Plateosaurus. *(Based on J. Sibbick)*

Apatosaurus. *(Based on G. Caselli)*

Ornithischian dinosaurs

The Ornithischia included animals that were more varied in appearance and adaptation than the Saurischia, and all were herbivorous in habit. There are some late Triassic ornithischian remains known from South Africa, but the main development and evolution of the group occurred during the Jurassic and Cretaceous. Ornithischians can be divided into the ornithopods (bird-footed dinosaurs), the stegosaurs (plated dinosaurs), the ankylosaurs (armoured dinosaurs), the ceratopians (horned dinosaurs), and the pachycephalosaurs (solid-skulled dinosaurs).

Fabrosaurs (= Fabre's reptile) and heterodontosaurs (= different tooth reptiles) were two groups of the smallest ornithopods, ranging in size from only ninety to 135 centimetres. They were agile, bipedal plant-eaters whose remains have been found in Jurassic deposits of North America, Africa, Europe and Asia. Representative genera included *Lesothosaurus, Heterodontosaurus, Scutellosaurus* and *Fabrosaurus*.

In middle Jurassic to late Cretaceous deposits are found the remains of larger ornithopods, the hypsilophodontids (= crested tooth) which ranged from two to 6.5 metres in length. Since they had larger jaws than the fabrosaurs, they were able to store food in cheek pouches while chewing plant material. Some hypsilophodontids were previously thought to be arboreal tree-dwelling animals, but this is now known to be incorrect and they were probably quite fleet-footed. The group included both bipedal and quadrupedal forms and remains have been found in North and South America, Africa, Europe and Australia. *Hypsilophodon, Tenontosaurus* and *Dryosaurus* were well known genera.

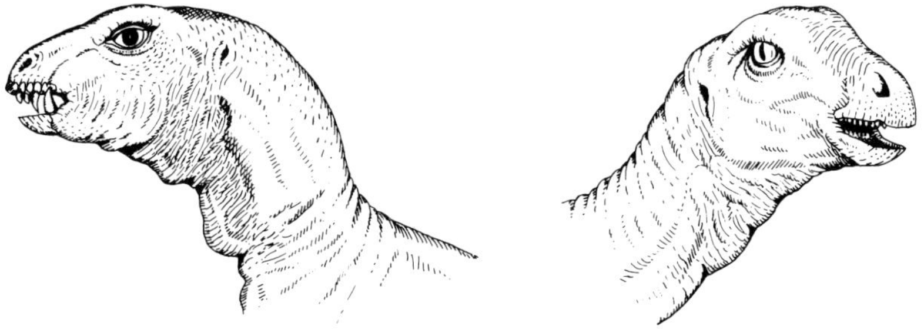

Left: Heterodontosaurus. Right: Hypsilophodon. *(Based on J. Sibbick)*

The iguanodontids (= iguana tooth), the first of the large-bodied ornithopods that evolved, are one of the best known ornithischian groups. From five to ten metres in length, these were medium to large, quite heavily-built bipedal dinosaurs whose remains have been found in deposits from the late Jurassic to the late Cretaceous. Iguanodontids had large, horse-like snouts, many grinding teeth, and broad toothless beaks for cropping foliage. A curious feature of the anatomy was the upward-pointed thumb "spike" that may have been used defensively by these heavy and cumbersome animals. Genera included *Iguanodon, Muttaburrasaurus, Camptosaurus,* and *Ouranosaurus,* and remains are known from Asia, North America, Europe, Africa and Australia.

The last and most successful group of ornithopods to develop were the hadrosaurs (= bulky reptiles) which ranged from middle to late Cretaceous in age. During this time, they evolved into a remarkable variety of species. Although in terms of body shape there was little variation (apart from size – four to twelve metres in length), the various head crests of different species were very distinctive. In the jaws of hadrosaurs, a formidable battery of hundreds of teeth permitted an efficient chewing mechanism capable of coping with

Iguanodon. *(Based on Z. Burian)*

many types of herbaceous material. Some hadrosaurs are known to have tended to their young by feeding them within nests in communal nesting grounds. Hadrosaurs were common in North and South America, Europe and Asia, and representative genera included *Bactrosaurus, Anatosaurus, Corythosaurus, Parasaurolophus, Tsintaosaurus*, and *Saurolophus*.

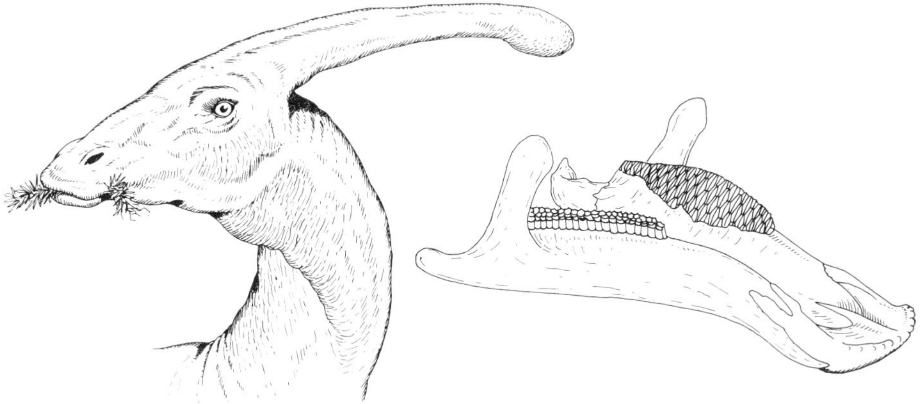

Left: Parasaurolophus. Right: *hadrosaur lower jaw and tooth batteries. (Based on G. Casselli)*

Of the non-ornithopods, the stegosaurs (= roof reptiles) were perhaps the most bizarre. Found from middle Jurassic to late Cretaceous deposits, stegosaur remains are known from North America, Asia, Africa, India and Europe. They were medium to large (2.5 – 7.5 metres long) heavily-built quadrupedal animals characterised by double rows of tall spines or wide plates (depending on the genus) running along the top of the back and tail, the end of which was also armed with large spikes. There is some evidence that the plates of some species acted in a thermoregulatory capacity, helping to cool or heat the animal as required. Stegosaurs typically had very small brains, which considering the size of these animals, were perhaps the smallest of any dinosaur. Genera included *Stegosaurus, Tuojiangosaurus*, and *Kentrosaurus*.

The Ankylosauria (= fused-together reptiles) were even more armoured than the stegosaurs, and this group included the club-tailed ankylosaurids and the nodosaurids which lacked tail clubs. Mostly of

Tuojiangosaurus. *(Based on J. Sibbick)*

Cretaceous age, and varying greatly in size from two to ten metres in length, ankylosaurs were the Mesozoic analogues of modern giant land-tortoises. They were very heavy, quadrupedal dinosaurs with low-slung, broad, armoured heads with the armour continuing to cover the entire back and tail, thus forming a type of fortified carapace. Most ankylosaurid remains come from Mongolia, with two genera also known from North America. Genera included *Ankylosaurus, Pinacosaurus, Saichania, Sauroplites, Shamosaurus, Talarurus*, and *Tarchia*, while nodosaurid genera (found only in European and North American deposits) included *Hylaeosaurus, Polacanthus* and *Nodosaurus*.

Pinacosaurus. *(Based on G. Olshevsky)*

The Ceratopia (= top horn-faced reptiles) contained the psittacosaurs and the protoceratopians as well as the larger, rhinoceros-like "true" horned dinosaurs. *Psittacosaurus* (= parrot reptile) was a small two-metre-long dinosaur from middle Cretaceous Gobi deposits. It was able to progress either bipedally or quadrupedally and had a short, parrot-faced skull, though the body showed ornithopod affinities. *Psittacosaurus* appears to show characteristics between the ornithopods and the ceratopians.

Protoceratopians were quadrupedal dinosaurs up to two metres in length from the late Cretaceous of Asia and North America. There were no horns on the faces of protoceratopians but they did have the well-developed neck-shields evident in the "true" horned dinosaurs. The best known protoceratopian is *Protoceratops* from Mongolia which is known from many complete skeletons and fossilised nests of eggs (chapter three). Like hadrosaurs, *Protoceratops* was a dinosaur that practised communal nesting. Other protoceratopian genera included *Bagaceratops*, *Leptoceratops*, and *Microceratops*.

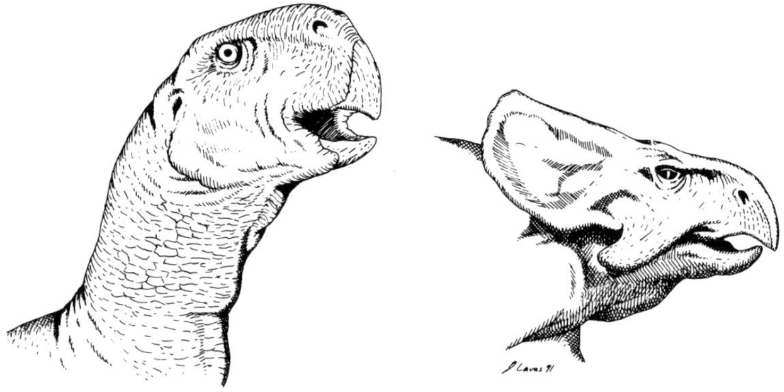

Left: Psittacosaurus. *(Based on J. Sibbick).* Right: Protoceratops. *(Based on Z. Burian)*

The "true" horned ceratopians were all large, quadrupedal dinosaurs from the late Cretaceous of North America, varying from five to nine metres in length. They are believed to have wandered the open plains in large herds, not unlike the modern bison. All had relatively short tails and hind legs longer and more stout than forelegs. Their very heavy heads bore a variety of horns on the snouts and upon the massive, bony neck-frills that covered the broad neck. The genus *Torosaurus* had the largest skull of any known land animal, and its 2.6 metre-long head was the size of a small automobile. Other well known genera included *Styracosaurus*, *Triceratops*, *Brachyceratops*, *Monoclonius*, *Chasmosaurus* and *Pentaceratops*.

Styracosaurus. *(Based on Z. Burian)*

The final group of ornithischians were the pachycephalosaurs (= thick-headed reptiles) that are mostly known from late Cretaceous deposits in Mongolia and North America. These unusual reptiles, which varied in size from one to eight metres, are a rather poorly known group of bipedal dinosaurs. They were notable for their exceptionally thick, dome-like skulls that endowed them with a misleading appearance of intelligence. It has been suggested that, as is the case with many modern ungulates, the male pachycephalosaurs may have engaged in competitive head-butting matches during the

Pachycephalosaurus. *(Based on J. Sibbick)*

mating seasons. Recently, new material found in the Gobi has provided much needed information on this group. Known genera included *Goyocephale*, *Homalocephale*, *Pachycephalosaurus* and *Stegoceras*.

1.3. The End of the Mesozoic

During the Triassic and Jurassic periods, the composition of dinosaur communities from many parts of the world showed little variation as there were few physical barriers such as major mountain chains or seaways to separate populations. The beginning of the Cretaceous period generally saw a continuation of Jurassic-type environmental conditions, with the continental landmasses still lying close to sea level and the climate remaining warm, humid and mild. The Cretaceous flora consisted of many of the older Jurassic varieties such as ginkgo, conifers, and sequoias, but this was also the time of the development of the flowering plants, the angiosperms, and many trees and shrubs would have appeared familiar to us. Firs, oaks, poplars, willows,

The remains of theropod dinosaurs are particularly common in the late Cretaceous Gobi strata. Best known of the large theropods is the carnosaur Tarbosaurus bataar *which was the longest of the known carnivorous dinosaurs and is believed to have been capable of considerable speed in spite of its great size. The species is known from many well preserved, complete skeletons representing both juveniles and adults. (Based partly on Z. Burian)*

birches, maples, sycamores, ash, figs, and breadfruit trees were all to be found, while the jungle of dense undergrowth included magnolia, cinnamon, dogwood, viburnum, laurel, polygonums, grape vines, lilies, heather and primulas. Social insects such as termites and bees evolved soon after the angiosperm radiation, so while bees and other insects dispersed pollen as they hovered between brightly-coloured blooms, termite colony mounds dotted the monotonous landscapes of many areas of rolling steppe.

In the mid-Cretaceous, about 100 million years ago, major tectonic upheavals resulted in the final break up of the continents. Australia and Antarctica began to part, and the subcontinent of India broke away from Africa and Australia and moved north, eventually to meet Asia and force up the Himalayan ranges. As Africa and South America had drifted apart, the South Atlantic seaway opened into a narrow, rift-like arm. The narrow land bridge connecting Europe to Africa was flooded over, and Europe had separated from North America.

After the splitting of the supercontinents had begun, and seas separated the different dinosaur communities, some dinosaurs began to evolve at an accelerated rate. This resulted in many new herbivorous ornithischians, such as duck-billed hadrosaurs and various large, horned ceratopians. And while most of the plated stegosaurs had already succumbed in the Jurassic, the Asiatic armoured ankylosaurs diversified widely during this time. Small and medium bird-like theropods also changed and evolved rapidly. Some previously wholly carnivorous coelurosaurs became omnivorous while giant theropods like the carnosaurs became ever larger and more common.

In the Gobi regions, the dinosaurs enjoyed their peak of diversity toward the end of the Cretaceous. This part of Asia had constituted the eastern section of the Laurasian supercontinent while western North America formed the western portion. While other continents continued to separate, the Bering Straits area between eastern Asia and western North America was also probably separated by marine barriers for much of the Cretaceous. And although some North American dinosaurs appear to show close resemblances to those of eastern Central Asia, many groups of Mongolian dinosaurs, including complete families, are not represented in North American deposits.

Toward the end of the Cretaceous, all dinosaurs from the Gobi as well as all other parts of the world began to decline, and dinosaur fossil remains disappear completely at or soon after the Mesozoic-Cenozoic strata boundary. Other reptile groups also vanish from the fossil record around this time, including the pterosaurs, the plesiosaurs, the ichthyosaurs, and the mosasaurs (though there is some evidence that the pterosaurs and ichthyosaurs became extinct somewhat earlier than the other groups). No single explanation appears to satisfactorily explain why these particular reptiles failed to make the Mesozoic-Cenozoic transition whilst many others such as the crocodilians, turtles and tortoises, lizards and snakes, and the rhynchocephalians (the tuatara of New Zealand), managed to survive. What is even more puzzling is that these extinctions took place during a period when many dinosaurs showed their greatest diversity. Several of the main theories that have been advanced at one time or another to account for dinosaur extinction should be briefly mentioned.

1) The theory of racial senescence, once popular in the late 19th century, assumed that the development of a race (or group of animals) followed the same natural laws as govern the development of an individual, which only has a fixed life span. It was believed that after millions of years, the dinosaurs may simply have outlived their allotted time span.

2) The theory of interclass competition, which considered that the early mammals could have contributed to the downfall of the dinosaurs by perhaps feeding on their unguarded eggs.

3) The celestial impact theory which includes evidence that a giant meteorite or asteroid (or similar body) plummeted into the Earth and disintegrated, producing dust clouds that spread about the globe and inhibited the photosynthetic processes of plants, thus causing firstly the downfall of the herbivores, and secondly, the carnivores that fed upon them.

4) The changing vegetation theory, which notes that the worldwide increase in angiosperms over the older Jurassic-type flora in the Cretaceous, produced a major dietary change for herbivorous dinosaurs. Many dinosaurs were thought to have not been able to cope with the new low oil content foliage, with their decline, as in theory 3), producing a follow-on effect upon the carnivores.

5) The theory of climatic change, which suggests that the cooler and more seasonal temperatures of the late Mesozoic produced a climatic regime to which all dinosaurs, being typically reptilian in physiology, could not adapt.

Although once popular, the racial senescence theory is no longer seriously considered by palaeontologists or biologists. Likewise, the theory of mammalian competition, originally proposed by the Georgian palaeontologist Davitashvili, is no longer supported. Although it is possible that some mammals ate some dinosaur eggs, mammals were certainly not a major ecological group during the Mesozoic, being in the main, very small animals of retiring and nocturnal habits. Apart from this, it is doubtful whether mammals would have even inhabited the poorly-drained, open coastal belts occupied by most types of dinosaur. They were far more likely to have sought upland areas with dense forest cover.

The celestial impact theory, whose evidence consists of high concentrations of iridium at the Cretaceous-Paleocene boundary of some sites, and which has been much discussed in recent years, cannot at the present time account for the fact that some groups of animals were selectively affected while others were not. Although certain plant groups did decline at about the Mesozoic-Cenozoic boundary, the extinctions tended to be geographically diverse, moderate in extent, and in many cases occurred after the dinosaurs had already disappeared. So on its own, this theory is somewhat inadequate.

The theory of climatic change deserves further consideration. In this context, the strata found in Gobi and other continental areas are important in that they exhibit fossiliferous sequences of late Cretaceous beds containing both dinosaur and mammalian remains together. In some cases, as in some of the North American sequences, these can be directly compared with early Cenozoic (Paleocene) outcrops containing only mammalian fossils, and some insights can be gained as to the faunal and geological changes that occurred during this time. It is known, for example, that certain types of continental deposits characteristic of Mesozoic strata are no longer present in the Cenozoic strata above. Mesozoic terrains included few mountain ranges or areas of greatly irregular, elevated topography. Most continental terrain consisted of broad, gently undulating lowlands with shallow, flooded basins and extensive marshlands. Such a homogeneous environment formed an ideal habitat for large-bodied animals like dinosaurs. Climatic conditions were also equitable and mild, well suited to all forms of reptilian life. Since the Jurassic, the continents had begun to move apart, and when Gondwanaland started to break up, world sea levels rose markedly. The result was that warm, shallow seas encroached over extensive areas of the low lying continents. Due to the moderating influence of large bodies of water, climatic conditions over most of the globe were subtropical and humid, with little seasonal temperature variation.

By the end of the Cretaceous, however, the continental plate movements were accelerating, raising large sections of land and forcing the uplift of mountain ranges at the conflicting plate boundaries. Many dinosaurs would have found their once equitable domains becoming progressively drained and fragmented. Some dinosaurs would have been more affected by such changes than others; for example, some sauropods and hadrosaurs appear to have been at least partly aquatic in habits, and were thus more directly affected than the armoured and horned dinosaurs that were more likely to have been steppe or plateau dwellers.

The new continental positions also resulted in the development of cooler polar regions and more seasonal temperature variations, conditions that today disadvantage poikilothermic reptiles living in high latitudes. This raises the question of dinosaur physiology, and whether they were poikilothermic and relatively dependent on external temperature consistency, or homoiothermic and more independent of it. Some scientists, notably Armand de Ricqles, have examined the bone histology of archosaurs and other reptiles and amphibians, and found densely vascularised bone in the archosaurs as well as in therapsids. This condition is generally associated with high metabolic rates and a homoiothermic physiology, as is characteristic of mammals and birds. However, such bone tissue is absent from some smaller species of mammals as well as being present in some undoubted poikilotherms such as turtles and crocodilians. As the evidence stands, the only conclusion to this question is that the coelurosaurs were probably the only group to which most of the evidence of homoiothermy is applicable. Studies of Komodo dragons (*Varanus komodoensis*) have shown that reptilian body tissue has quite low heat conductances in any case, and in reptiles of body weight over 100 kilograms, the temperature differentials actually begin to approach those of mammals of similar size. Many of the larger dinosaurs, with very low surface-to-volume ratios, did not need to be homoiothermic and may have had greater difficulties losing heat in the summer rather than maintaining a high enough working temperature in winter. So it is reasonable to assume that topographic changes were of greater consequence, at least initially, to all medium and large dinosaurs.

Another key question concerns the viability of dinosaur eggs. Study of many dinosaur eggshells preserved in the French Pyrenees show that those from early Mesozoic strata are often up to 2.5 millimetres thick, which is what one would expect for eggs of this size. But shells from late strata appear to get progressively thinner until at the very latest layer, they are only one millimetre or so. Birds have been known to lay very thin eggshell under stress, such as a prolonged cold period for example. If the shell is too thin, the embryos will not develop because they are unable to absorb enough calcium to complete the building of their skeletons. Temperatures may have had other detrimental effects as well. Nests from several groups of herbivorous dinosaurs are now known, and these often show that the eggs were laid in one horizontal layer within a shallow depression. Such an arrangement would have been suitable for non-seasonal climates such as existed throughout much of the Mesozoic, but if the incubation periods were as long as they are for some modern reptiles, the cooler months that developed towards the late Cretaceous might have proven fatal to dinosaur embryos. Of the many herbivorous dinosaur nests that have been found in late Cretaceous deposits of the Gobi, large numbers contain unhatched eggs still in the same positions as they were laid. Perhaps if the incubation periods of herbivorous dinosaur eggs had been shorter, their fate may well have been different.

The end of the Mesozoic also brought about an increase in stormy and unsettled weather conditions that accompanied the new pattern of continental oceanic current flows. The delicately-boned pterosaurs, with their thin, large-surface area wings so well suited to cruising in the calm Jurassic and early Cretaceous skies, could no longer survive. And the draining of the huge inland seas of many areas affected different forms of floating plankton, causing the extinction of the calcareous (chalky) varieties as well as other invertebrates, such as the cephalopod ammonites (shelled relatives of modern squids and cuttlefish). The primary herbivores which depended on plankton would have declined as a consequence, while the giant marine reptiles that likewise fed upon the herbivores (and the ammonites) also lost their food source. With the two orders of dinosaurs as well as the pterosaurs now gone, the only surviving archosaurs were the crocodilians, who for some reason, managed to evade extinction. Consequently, they have come through the last 200 million years or so largely unchanged. Living in the tropical latitudes as aquatic predators and scavengers, they are living representatives of a once far greater reptilian fauna from the Age of Reptiles.

1.4. Dinosaurs and Birds

Since the 1970s, there has been a considerable amount of evidence presented to suggest that dinosaurs did not all become extinct at the close of the Mesozoic. They may still be singing outside our windows on summer mornings, foraging for worms on the front lawn, and roosting in trees in the back yard. What is the evolutionary relationship between birds and dinosaurs? This question has been actively debated in recent years, but there does not yet appear to be a conclusive answer that accommodates all the known facts in the case.

The most celebrated avian fossil is of course that of *Archaeopteryx*. Indeed, it has often been described as the single most important fossil find ever. The saga of the identification of *Archaeopteryx* itself unfolds like a mystery novel worthy of the best fiction writer. The story begins in the 19th century in the limestone quarries of Solnhofen in southern Germany. Here, the stone is of late Jurassic age, and is of such a fine grain that it was used in the printing industry. In 1855, a small, fragmentary and inconspicuous reptilian skeleton was exposed in the limestone. In fact, this was the first example of *Archaeopteryx* to be found, but as there was nothing unusual about it (there were no feather impressions) it was not recognised as being of any great value. This find eventually ended up in the Teyler Stitching Museum in the Netherlands labelled as a pterosaur, and it was not until some 115 years later, in 1970, that John Ostrom of Yale University recognised its true identity.

In 1861, a much better preserved example was found by Dr. Karl Haberlein, and this more complete skeleton clearly showed fine impressions of bird-like feathers around the animal's forearms and long, reptilian tail. Dr. Hermann von Meyer named this new find *Archaeopteryx lithographica*, meaning "ancient wing from the

Archaeopteryx lithographica. *(Based on Z. Burian)*

lithographic stone". The British Museum eventually purchased the fossil for the then staggering figure of £500, almost two years worth of the Museum's total budget.

Coming as it did just two years after the publication of Charles Darwin's "Origin of Species by Means of Natural Selection", the discovery of the 1861 *Archaeopteryx* provided Darwin's supporters with a convenient example of a "missing link" between two vertebrate classes, the Reptilia and the Aves. Because the fossil record was known to be so incomplete, Darwin had envisaged that many such transitionary forms must have existed in the past, each exhibiting characteristics both from its ancestors as well as its descendants. Thomas H. Huxley, Darwin's ardent supporter, described the new *Archaeopteryx* as the ideal evolutionary link between modern birds and the dinosaurs (of which only a few species had then been described). This argument has more recently been reconsidered in the light of new evidence, as will be discussed later.

1877 saw an even better example of *Archaeopteryx* found in the Solnhofen limestone, this time by Karl Haberlein's son Ernst. This is undoubtedly the finest specimen ever found, and the German government was petitioned against allowing the precious relic to leave the country. A German industrialist eventually purchased the fossil and later sold it to the Berlin University, this time for the equivalent of no less than £1000. The specimen is now housed in Berlin's Von Humboldt Museum. A further *Archaeopteryx* was uncovered in 1956, and in 1973 Peter Wellnhofer identified yet another example, the so called "Eichstatt" specimen that had been labelled as a coelurosaurian dinosaur since its discovery in 1951. In 1988, yet another previously unidentified specimen turned up in a private collection belonging to a former mayor of Solnhofen. This was recognized as an *Archaeopteryx* by the director of the Jura-Museum in Eichstatt, and is the largest specimen yet found.

Here it is necessary to reconsider the basic classification of the archosaurs to which *Archaeopteryx* is related. As stated previously, the Thecodontia or "socket-toothed reptiles" constituted the basic archosaur stock that gave rise to the other extinct groups, that is the dinosaurs and the pterosaurs, as well as the still-surviving crocodiles (and related types). The term "Dinosauria", meaning "terrible lizards", was originally introduced in 1842 by the British anatomist Sir Richard Owen (1804-1892) to include the then several known types of unusually large, extinct reptiles. Owen's classification stood unaltered until 1877 when Harry Govier Seeley, Professor of physical geography at Bedford College, London University, advanced a more comprehensive grouping involving the two currently recognised orders. Dinosaurs with the pubic bones directed anteroventrally (forwards and downwards) as in

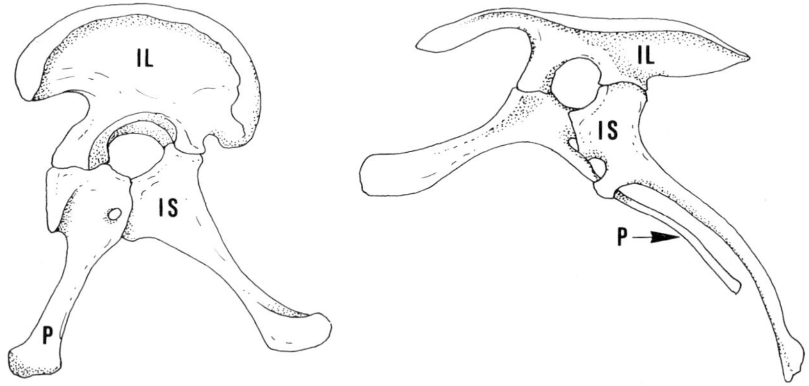

Comparative structure of the pelvis of a saurischian dinosaur (left) *and an ornithischian dinosaur IL = ilium IS = ischium P = pubis (Based on Colbert)*

typical reptiles, were assigned to the order Saurischia. Those with a postpubic process (the supposed homologue of the saurischian pubis) directed posteroventrally (backwards and downwards), as in modern birds, were assigned to the order Ornithischia. Seeley included both orders in the subclass Archosauria along with the Crocodilia, Pterosauria, and the Thecodontia.

In 1926, Gerhard Heilmann studied the anatomy of *Archaeopteryx* in great detail, comparing it to many other better known types of reptiles. He concluded that the small saurischian coelurosaurian dinosaurs were probably the closest relatives except for one aspect of their anatomy, their apparent lack of clavicles (collar bones or "wish bones" as they are known in birds). As a result, Heilmann proposed that *Archaeopteryx* evolved from a much more ancestral type of archosaur that had firstly developed tree-dwelling habits and later evolved feathers as it glided between branches.

Heilmann's views stood largely unchallenged until 1970, when Peter Galton proposed the ornithischian dinosaurs as the ancestral group for the birds. Although the preservation of *Archaeopteryx* does appear to show an ornithischian-type pelvis, other aspects of its anatomy are very difficult to reconcile with that of the ornithischians, and Peter Galton has since concluded that the theory is incorrect. At about the same time (1972), Alick D. Walker advanced the crocodile origin theory. He claimed that birds and crocodiles shared important aspects of skull morphology and heart structure. Some support for this theory came from L.D. Martin and K.N. Whetstone but Walker himself eventually found the theory inadequate and has since abandoned it.

Considerable backing has come for John Ostrom's saurischian origin theory that was proposed in 1973. It is basically a revised version of the argument taken by T.H. Huxley but backed up by more detailed evidence. Ostrom has listed some twenty anatomical features that he claims are shared by *Archaeopteryx* and coelurosaurs, ranging from the three-fingered hand, the morphology of the fingers, the four-toed foot, the forelimb and pectoral girdle structure, and the pelvic arrangement. While at first glance, this last factor may seem incongruous, Ostrom claimed that the pubic bone as preserved in *Archaeopteryx* had been dislocated and altered during preservation. He believes therefore that its original direction lay forwards and downwards as in saurischian dinosaurs. The argument has been further complicated because the diagnostic value of pelvic arrangement in dinosaurs has recently been questioned. In 1979, the Mongolian palaeontologist Altangerel Perle described an unusual new group of theropods found in the Gobi. Classified in their own infraorder, the Segnosauria (see chapter five), these saurischian dinosaurs are unique in that they have pubic bones directed downwards and backwards as in the ornithischians.

However, Ostrom was more confidently able to list the similarity between the pectoral girdles of *Archaeopteryx* and saurischians. By that stage, it was known that contrary to what Heilmann had believed, some saurischian dinosaurs did in fact possess clavicles. A joint Soviet-Mongolian Expedition to Mongolia in 1972 produced several complete skeletons of oviraptorosaurs. These skeletons show clearly-preserved collar bones fused together in a similar fashion to the "wish-bones" of birds. However, as is the case with the pelvic girdle, this evidence now also appears to be inconclusive. Recent embryological work suggests that these dinosaur clavicles may not even be the same bones as their avian counterparts, and it may still be simply a case of convergent evolution between birds and saurischians.

Opposition to the saurischian origin theory has come primarily from the work of Max Hecht and Sam Tarsitano who favour a thecodont origin for the birds. They note that the theropod dinosaurs had already lost the fourth digit of the hand, whereas embryological studies indicate that this digit is still retained in modern birds. They also claim that some characters listed by Ostrom as evidence for a saurischian origin are ancestral traits that would be expected to be shared between the two groups even if they were not closely related. Hecht and Tarsitano propose, as did Heilmann, that *Archaeopteryx* evolved from a pseudosuchian thecodont at a very early stage of archosaur evolution.

A small fossil found in 1970 by the Russian palaeontologist A.G. Sharov in Kirgizia in Central Asia has added yet another interesting aspect to this argument. Found in early Triassic strata, *Longisquama insignis* was a tiny pseudosuchian thecodont whose body was covered in long overlapping scales which along the back developed into lengthened spatula-like appendages two to three times longer than the rest of the body. These appear to have functioned as a parachute to help break the animal's fall as it glided from tree to tree. The elongated scales on the back of the forelimbs are thought to represent an early stage in the development of feathers, and a set of avian-like clavicles are clearly preserved. Though *Longisquama* itself is not considered to

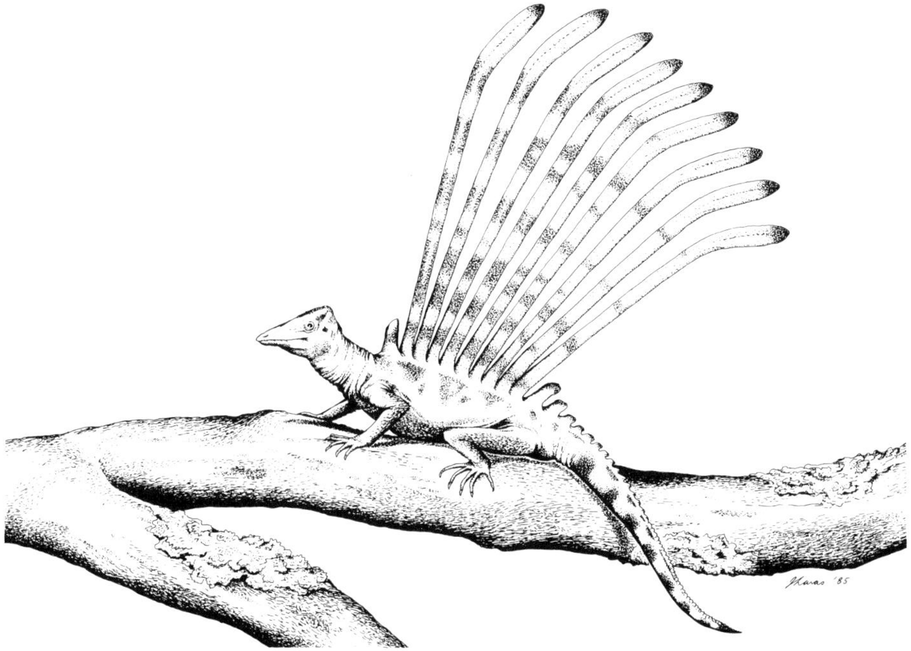

The pseudosuchian thecodont Longisquama insignis, *which lived about 80 million years before* Archaeopteryx, *is thought to have glided between branches and trees using its long dorsal scales as a parachute. Length not including tail was fifty millimetres.* (Based on Sharov)

be ancestral to modern birds, it does show to what extent some early thecodonts were experimenting with passive flight at a far earlier stage than *Archaeopteryx*.

Perhaps not surprisingly, other avenues of research into the question of bird-archosaur relationships also prove inconclusive. Studies on dinosaur eggshell microstructure have been carried out in recent years, mostly on material recovered from Mongolia, and from Provence in France. Researchers have found that eggshell from modern birds often shows closer morphological similarities to those derived from ornithischian dinosaurs (protoceratopians and hadrosaurs) than it does to eggshell believed to be derived from saurischian dinosaurs (sauropods). However, such results may indicate that due to reasons of ecology or life-habits, the ornithischians developed more of a bird-like shell than did the saurischians, even though the saurischians were more closely related.

If the question of the origin of birds appears confusing, then the closely-related question of the origin of bird flight is less so. Basically, the attainment of avian flight has been explained by one of two theories. The first, the cursorial theory, was initially proposed by Samuel W. Williston in 1879, and expanded by Baron Franz Nopcsa in 1907 and 1923. This theory envisages an early fast-moving bipedal "proto-avian" that developed enlarged scales or feathers on the forearms. Flapping these "proto wings" was combined with progressive leaps from the ground in an effort to increase speed. The cursorial theory has not been well supported because there is no aerodynamic evidence that such forearm activity would be able to add to an animal's velocity.

More recently, in 1974, John Ostrom introduced a variation of this theory which has been called "the cursorial predator theory" or more commonly "the insect net hypothesis". Ostrom described *Archaeopteryx* as a feathered coelurosaur with flight being attained as a result of predaceous habits. Enlarged contour feathers on the forelimbs might have been used as a trapping device into which flying insects were corralled before being grasped by the beak or hands. Ostrom suggests that the morphological similarity between the hand, arm and pectoral girdle of *Archaeopteryx* and theropods like *Ornitholestes* and *Velociraptor* was due to similar predatory functions rather than to convergent evolution. He believes that feathers existed in pre-*Archaeopteryx* forms and arose in connection with the need to control heat loss, or as heat shields against heat gain in hot climates. Thus a state of warm-bloodedness would have preceded the attainment of flight. Some authorities, principally Alan Feduccia, and Tarsitano and Hecht, have criticised the insect net hypothesis claiming that the aerodynamic nature of feathers would cause air pressure to expel rather than contain any insects when the forearms were used in this fashion.

The other major theory of bird flight, commonly called "the arboreal theory" has received much more support than the cursorial theory. Originally proposed by Othniel Charles Marsh in 1880, the theory was later expanded by Heilmann during his work on the origin of birds. Early proto-birds are thought to have first become arboreal in habits, later using their feathered- limbs and tails to help them glide

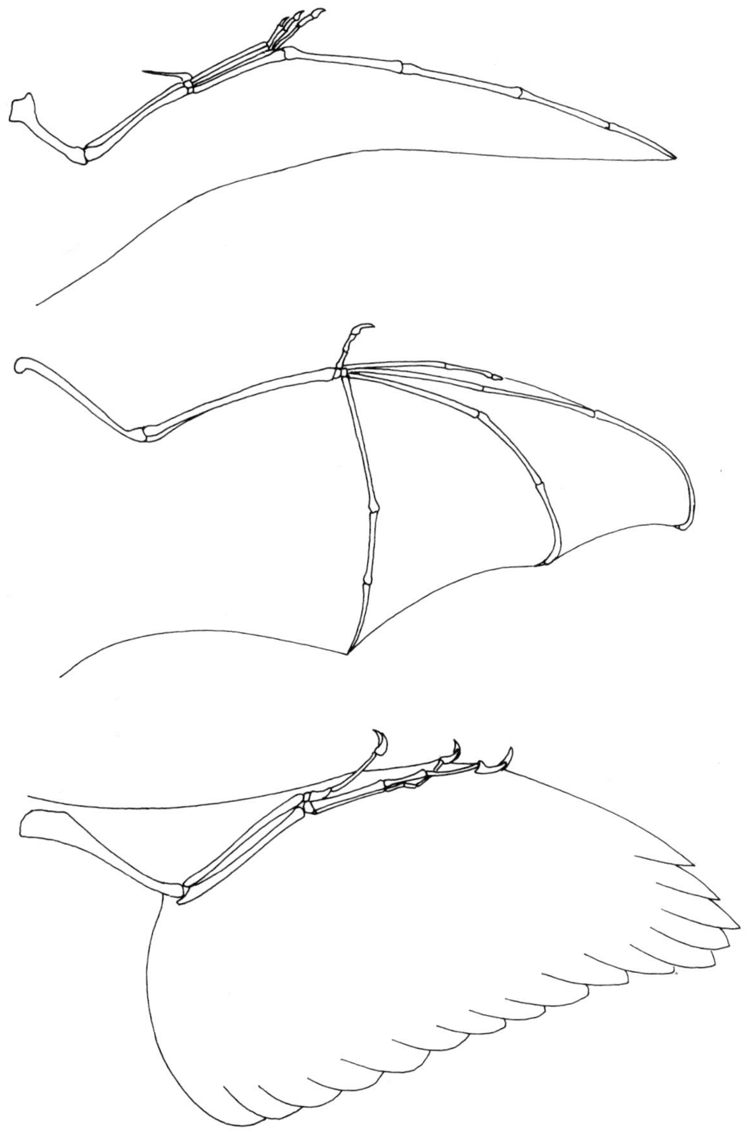

Schematic views of the wings of flying vertebrates. Top: *pterosaur.* Middle: *bat.* Bottom: Archaeopteryx. *Note that in the wings of both pterosaurs and bats, the structural integrity of the wing membrane is aided by supporting collagen fibres. (Pterosaur and bat based on Padian,* Archaeopteryx *based on Yalden)*

from tree to tree or to help them break their fall as they descended to the ground. Eventually, active flight developed with modifications of the pectoral girdle and associated muscles.

Some vertebrates today employ gliding as a means of moving from tree to tree. These include various species of frogs (genus *Rhacophorus*), lizards (genus *Draco*) and flying squirrels (genus *Pteromys*). Apart from *Draco* in which a fold of skin known as a patagium is stretched between outstretched ribs, all these gliding forms use a flight membrane stretched between the fore and hindlimbs. In proto-avians, feathers would have served the same purpose. In 1915, Walter Bock offered a sequence of stages in the evolution of avian flight. This was 1) an ancestral ground-dwelling quadrupedal reptile, 2) a bipedal ground-dweller, 3) an arboreal biped, 4) leaping between closely spaced trees, 5) parachuting, 6) gliding longer distances, and 7) active powered flight.

There are no complete fossil remains of such transitional stages in birds, but there are for the Mesozoic masters of the skies, the pterosaurs (with the difference that the immediate ancestors of pterosaurs were quadrupedal rather than bipedal). After the discovery of the first pterosaur remains in Bavarian limestone deposits in 1784, the physiology of these flying reptiles had evaded explanation because it could not be shown how a poikilothermic reptile would be able to produce the energy demanded by active flight. Recent work by Kevin Padian has shown that these reptiles were active fliers with high metabolic rates similar to those of birds. The exact nature of the body insulation that permitted such metabolism was revealed in 1970 by one of A.G. Sharov's finds in Upper Jurassic rocks in Kazakhstan. Named *Sordes pilosus* (= hairy devil), this small pterosaur was completely covered in fine, fleece-like hair not only over the body but over the wing membranes as well. Pterosaurs were obviously highly active and successful flying archosaurs that were not only related to birds, but equalled their aerial achievements some 70 million years before *Archaeopteryx*.

In Lower Triassic rocks of nearby Kirgizia, Sharov also found another very important find. This was a tiny pseudosuchian thecodont that is believed to be the gliding ancestor of the pterosaurs. Originally named *Podopteryx mirabilis* (= marvellous foot-wing), this fossil has since been renamed *Sharovipteryx* (= Sharov wing) because the name *Podopteryx* had actually been given to a genus of fish 100 years earlier. *Sharovipteryx* was a tiny quadrupedal reptile with membranes of skin stretched between the elbows and knees and between the upper leg and tail. These appear to have been elastic and the amount of lift generated was primarily controlled by the position of the hindlimb. Pterosaur evolution and the attainment of powered flight proceeded with the development of the front membrane which became attached to the long fourth finger, and the degeneration of the membrane between the hindlimbs.

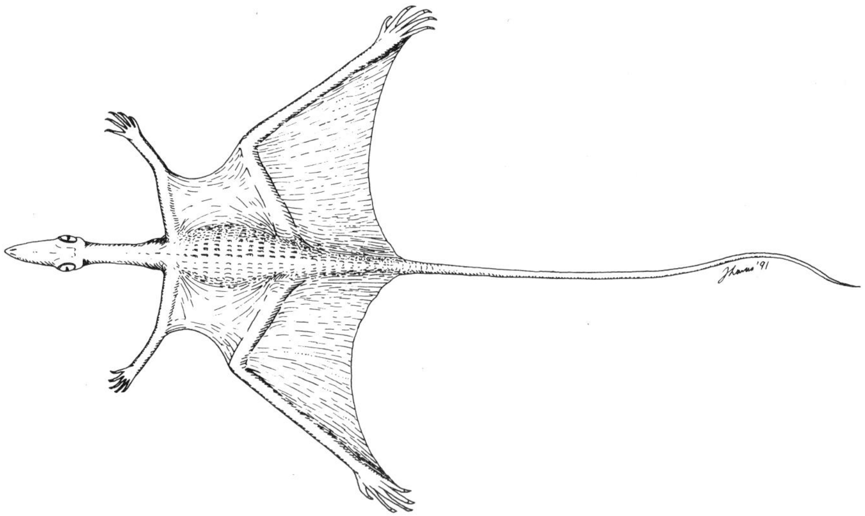

Sharovipteryx mirabilis, *a possible ancestor of the pterosaurs. Length not including tail was 100 millimetres. (Based on Sharov)*

There are several comparisons that can be made between the evolution of birds and pterosaurs. Like all post-*Archaeopteryx* birds, pterosaurs possessed pneumatic foramina in many postcranial bones (presumably to help cool the blood when body temperatures rose due to muscular activity). Other shared features of the skeleton include the acrocoracoid process, the deltopectoral crest, the broad, keeled sternum and braced coracoids. The wing mechanisms of birds and pterosaurs show interesting comparisons. In both groups, the wing was elevated by the powerful pectoral muscles which attached to the deep sternum. In modern birds, the upwards or recovery stroke is activated by the supracoracoideus muscle. This muscle operates via a tendon which runs along a reversed U-shaped course before it attaches to the humerus (upper arm bone). Consequently, the wing is elevated not by a dorsal muscle as one would logically expect, but by a ventral one. Kevin Padian has shown that this reversed muscle action also operated in the wings of pterosaurs.

In contrast, *Archaeopteryx* appears to have only had a weak cartilaginous sternum with no observable keel. The wings were probably lowered by the supracoracoideus rather than raised by it because this muscle had not yet reversed its action to become a wing elevator (the *Archaeopteryx* musculature is considered to be the more primitive condition), while the recovery stroke was probably powered by the relatively weak deltoid muscles. In fact, *Archaeopteryx* shows almost no skeletal features that can be directly compared to the flight apparatus of modern birds. By any analysis, it was little removed from being a feathered archosaur capable only of short bursts of flight. Palaeogeographical and palaeoecological evidence from the Solnhofen area shows that the habitat of *Archaeopteryx* was an area of coastal lagoons and small, semi-arid islands whose flora was low scrubby conifers. Within such vegetation, *Archaeopteryx* was able to climb and nest, and it probably lived a life not unlike that of a shorebird, scavenging on aquatic debris and perhaps diving upon fish and cephalopods near the water surface. In such an environment it only needed to fly from island to island, so there was little benefit in developing long-distance flight capabilities.

The exact nature of pre-*Archaeopteryx* types is very difficult to identify because of the similarities of their bones to those of some thecodonts and saurischian theropods. But as feathers are a major diagnostic feature of birds, another of Sharov's finds should be mentioned in this context. It is a single primary wing feather found in Upper Jurassic rocks, again in Kazakhstan. Described by Rautian in 1971, this feather is believed to have belonged to a proto-bird (named *Praeornis sharovi*) that occupied a pre-*Archaeopteryx* evolutionary position. In contrast to the feathers of all known ancestral birds (including *Archaeopteryx*), for which no major differences from those of modern birds can be discerned, the feather of *Praeornis* has a number of features which make it unique. The morphology of the bird feather consists of the rachis or main shaft from which branch the rami or barbs. Along the rami are series of comb-like barbules and these in turn are studded with hook-like hamuli (see diagram). The barbules and hamuli lock adjacent barbs together and evenly distribute the wind pressure over the feather surface during flight. Efficiency of flapping flight is enhanced by the barbs becoming thinner and lighter and by an increase in the number of barbs per unit length of the rachis. Modern birds have at least fifteen barbs per centimetre, *Archaeopteryx* has twenty-two, possibly because it was very heavy for its size compared to modern birds. The barbs of *Praeornis* however, did not exceed four per centimetre, and there were no barbules (and therefore no hamuli) at all. Rautian suggests that *Praeornis* was only capable of short bursts of energetic flight, perhaps to seize air-borne insects or to escape predators.

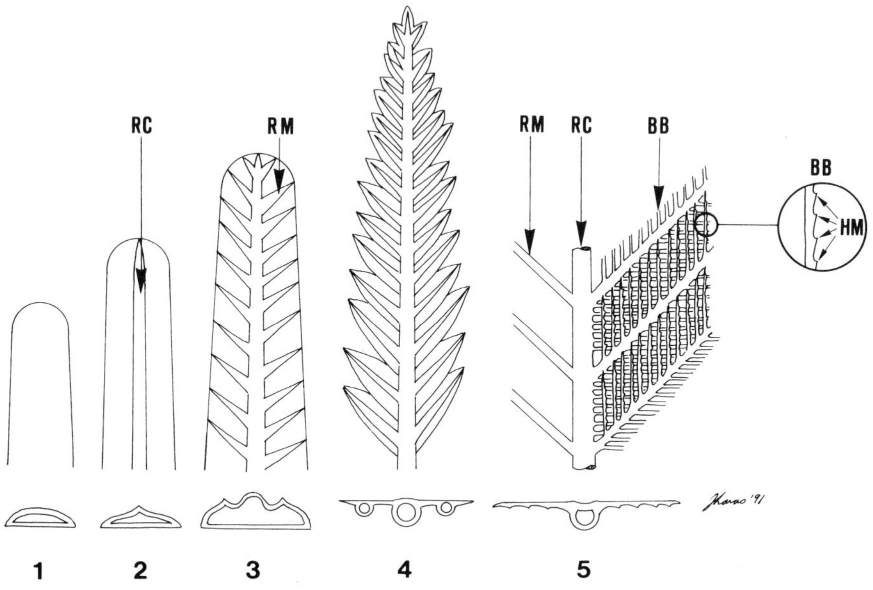

A sequence of possible evolutionary stages in the transition from a reptilian scale (1), through to the feather of a proto-avian such as Praeornis *(4), to the feather of a modern bird (5). Beneath stages 1-4 are shown transverse sections through the rachis, beneath stage 5 is a section through one of the rami. RC = rachis, RM = rami, BB = barbules, HM = hamuli. (Based on Rautian)*

As to the nature of post-*Archaeopteryx* forms that had not yet reached the evolutionary development of modern birds, a 1986 find from Las Hoyas in Spain is notable. This primitive bird lived well after *Archaeopteryx* in the early Cretaceous about 130-120 million years ago, and anatomically appears to be somewhat intermediate between *Archaeopteryx* and modern birds. Like *Archaeopteryx*, the structure of the pelvis, sacrum (lower spine), and hindlimbs show ancestral rather than well developed features, but it also shows some important skeletal flight adaptations. The tail vertebrae were fused together for in-flight stability and the collarbone was braced firmly to the breastbone, strengthening the shoulders during wing strokes.

Obviously, our knowledge of the evolution of birds is very incomplete, consisting mainly of fragmentary remains that represent isolated, transitionary stages. Many of the recent finds in this field have been made in Central Asia and in the Gobi where the remains of theropod dinosaurs are particularly abundant. In conclusion, it appears that the origin of birds lies either with the small saurischian theropod dinosaurs, or earlier back with the smaller pseudosuchian thecodonts, some of which became airborne in the early Triassic long before the appearance of the birds. Feathers most likely arose as a result of increased activity and a high metabolic rate, and later might have been used when arboreal proto-avians glided between trees in the Jurassic forests. At this time, the exact sequence of evolutionary stages leading from a ground-dwelling bipedal reptile to one capable of powered flight remains unknown. It is likely, however, that future palaeontological exploration in places like the Gobi will eventually fill many such gaps in our knowledge of the most accomplished aerial vertebrates ever to evolve.

Chapter Two

Philosophy, Geology, Palaeontology

2.1. Philosophers and Fossils

EVERYTHING now known about dinosaurs and other extinct animals comes from studying and understanding their fossilised remains, (the science of palaeontology), comparisons with surviving related types such as crocodiles and birds in the case of dinosaurs, and a knowledge of the geological and climatic conditions present at the time of fossilisation.

Palaeontology is one of the more recently developed sciences, there having been less than 300 years since it became widely accepted that fossils originated from once living animals and plants. Yet as early as 600 years B.C., some of the noted Greek philosophers had already begun to understand the meaning of fossils. Sea shells and the remains of other petrified marine creatures were frequently dug up in quarries that were many kilometres from the ocean, and such enigmatic objects naturally attracted the attention of the philosophical Greeks. The earliest writers to give their views on the topic were the philosopher Xenophanes of Colophon and the Lydian chronicler, Xanthos of Sardes. The original works of Xenophanes have unfortunately mostly been lost, but later writers recorded that he had studied fossil molluscan remains in mountainous areas far inland, coming to the conclusion that such areas must have at one time been inundated by the sea. For his part, Xanthos observed many fossilised shells in the interior of Asia Minor and recognised them as having once been living marine creatures. He also proposed that older remains should lie beneath those of more recent age, that is they will lie deeper as long as they have not been disturbed. This view was also suggested by Theophrastus of Lesbos, and today is recognised as a basic principle of stratigraphy.

The notable Greek historian Herodotus, from Halicarnassus, recorded the existence of sea shells in elevated areas of Egypt, again, many kilometres from the ocean. This, in addition to the shape of the Nile river delta system and the types of soils characteristic of that area, lead him to deduce that the ancient seas had once covered large areas of Northern Africa. These early insights were not limited to palaeontology either, for some original evolutionary views emerged at the same time. After viewing petrified fish within very deep strata, Anaximander of Miletus reasoned that fish must therefore be ancestral to all other backboned animals, including humans. The later writers of the Roman Empire expressed much the same ideas as their Greek predecessors, and Pliny, among others, suggested that primeval floods had greatly modified the areas now lying above sea level.

By the Middle Ages, naturalists and philosophers alike offered a variety of interesting but unscientific explanations when the odd fossil was dug up. Apparently, any correct interpretations gained by the perceptive Greeks were either unknown or dismissed. The most commonly expressed view, by far, was the theory of *vis plastica*, originally suggested by Aristotle (383-322 B.C.) and later perpetuated by his Arabic commentator, Avicenna (980-1037 A.D.). *Vis plastica* was described as a type of "fertile wind" of a creative and mysterious force of nature which sometimes produced the likenesses of organic things out of non-living matter. By this time, most people were prepared to view fossils as simple "chance" creations, sports of nature with no real significance. Coupled with this belief was a general acceptance that certain forms of "low" life, such as flies for example, were produced directly from dung heaps, while rats were thought to grow straight from carcasses. Given this, no great feats of imagination were required to accept *vis plastica* as a viable explanation, and the theory managed to survive largely intact well into the 18th century.

During the Renaissance, some people reached a truer understanding of fossils, in particular, the great inventor and artist, Leonardo da Vinci of Florence (1452-1519). His works on the subject are regarded as the beginnings of a systematic and scientific approach to the study of

Aristotle (383-322 B.C.).

Leonardo da Vinci (1452-1519).

palaeontology. Leonardo came to Milan in northern Italy in 1484, at the request of the Duke Lodovico Sforza, to give assistance with engineering work being carried out during the construction of that city's canal system. It was only a matter of time before he began to direct his attention to the geology at the exposed construction sites. Countless fossilised snails, mussels, oysters, crabs and fish were constantly turning up at the diggings. Nobody had been able to offer a reasonable explanation for why they should be there, as the sites were nowhere near the ocean. To Leonardo, who was already an accomplished physicist and mathematician, as well as having excelled in many other sciences and arts, the problem could only be approached in a scientific and methodical manner.

He immediately dismissed the theory of *vis plastica* to explain the petrified remains. To him, the whole idea must have appeared totally unacceptable, though he was at the same time wary of publicly denouncing a viewpoint that was widely held throughout the Western world. Tactfully, he systematically recorded all of his observations, finally arriving at the conclusion that these land areas were once submerged, during which time the various marine organisms had become trapped in the mud. Arriving at the same conclusions as the ancient Greek and Roman writers, he argued that the shells of the molluscs must have first filled with mud, then much later become petrified. The fish, he explained, had left hollows or impressions in the mud after death, these later having filled with different types of mud. The exact form of the body would become preserved in the older type of mud and remain so after petrification.

Leonardo was also critical of another widely accepted school of thought that was to become increasingly influential in palaeontological work in years to come. This was the theory of the Great Deluge, the world-wide flood, whose proponents were referred to as "diluvians". Like Leonardo, diluvians rejected the *vis plastica* concept, and accepted fossils as being the true remains of animals and plants. But in their minds, all fossils were remnants of victims of the world-wide flood described in the biblical *Book of Genesis*. That is, they were the remains of creatures that had all perished at the same time.

The first historical writings mentioning the Deluge go back to between 1000 and 2000 years B.C. and can be traced to an Assyrian document which was later (600 B.C.) translated into Babylonian. It tells of a Babylonian King, Xisuthrus, who was warned of an impending flood of great magnitude. The king and his closest friends constructed a ship which rode out the flood, later grounding upon a mountainside. Another version of the same story relates how the rains destroyed all life upon the Earth, except for the people and animals taken aboard the ship. There is some disagreement as to whether these descriptions refer to a universal flood or one which only affected local coastal areas, such as the Mesopotamian region. Inscriptions from the archaic Egyptian age and records from ancient Greece also mention floods of great magnitude, and in all cases, the waters were responsible for much loss of life.

It was in the 13th century that the suggestion of using the legendary deluge as an explanation for the occurrence of fossils was made by Ristoro d'Arezzo of Italy. D'Arezzo had begun his career as a naturalist and an ardent supporter of Aristotle's *vis plastica*, but after finding some very convincing fossilised snails and fish on a mountainside, he became disillusioned with the theory. The next best explanation for such anomalies was that the creatures had become caught in an untimely flood that had engulfed all the continents. Although few people took much notice of his writings at the time, by the 17th century, the deluge theory was revived and quickly gained widespread acceptance. By then, most Western theologians and academics were prepared to vigorously defend it.

In England, Thomas Burnet, a prominent writer on religious affairs, wrote extensively on the creation of the world and the effects of the Deluge. Burnet proposed that the world had been created by the Supreme Being as a kind of global stage for human activity. He imagined that in the early stages, it was perfectly spherical and uniform, a paradise (if somewhat monotonous) of continuous, unchanging climates. As it began to dry and solidify, huge cracks released waters that were previously contained within it, thus causing a devastating flood which all but destroyed the "first" world, leaving its surface scarred and rugged as it is today. Such ideas were not original though, and regarding the world as a planet in the process of decline from an original state of perfection is a concept often encountered in ancient cosmologies.

Some of Burnet's contemporaries began to speculate on the prehistory of the world using the evidence of the rocks themselves. Collecting fossils, or "figure stones" as they were often referred to, was becoming a popular pastime amongst scholars. At the same time, the debate on the nature of fossils was becoming more intense. Unfortunately, Leonardo da Vinci's excellent explanations of marine fossils had not been widely read. His collected works of more than 5000 pages of journals and notebooks were not published until many years after his death, and most people were not able to grasp the wealth of information that they contained. In England, the *vis plastica* concept continued to be perpetuated by such people as Edward Lhuyd and Martin Lister.

2.2. Against the Deluge

The most notable person to emerge in opposition to *vis plastica* was the great physicist Robert Hooke (1635-1703), who worked for the Royal Society in its early years. Hooke had a perceptive understanding of fossils and the sediments which contained them, and he concluded that petrifications were unquestionably the remnants of ancient life. How could a "fertile wind" be responsible for producing some of the very finely detailed fossils that had been found? Hooke realised that the oceans and land areas were constantly changing, never static. The flooding of low lying areas allowed the oceans to move in and erode large sections of land, eventually levelling out entire mountains, while elsewhere, volcanic eruptions and earthquakes could force open valleys and raise mountain chains out of level plains. Thus the Earth was in a constant state of subsidence and elevation, an endless conflict between the inexhaustible elemental forces.

Hooke rejected the concept that a single universal deluge could have accounted for such changes in the Earth's topography. Such an event would be too brief and transient. He decided that the most important factors must be the powerful earthquakes and volcanic eruptions that occurred from time to time. He also knew that the only way to provide evidence for his ideas was to find some historical event that recorded a great earthquake or eruption that had resulted in major geographical alterations, so he turned to Plato's account of the disappearance of that legendary island-continent ... Atlantis.

Atlantis was said to have been a continent lying against the Pillars of Hercules (two great rock formations in the Straits of Gibraltar). Greater in extent than Libya and Asia combined, it formed an empire that was a passage to the Mediterranean, reaching to Egypt and Tyrrhenia. It is said that the might of Atlantis was pitted against Egypt, Hellas, and all the surrounding Mediterranean countries. But the Athenians courageously repelled the Atlanteans and soon afterwards, a great earthquake struck the continent, and it disappeared beneath the waves. This was the explanation given for the extensive shallows found in that part of the Atlantic Ocean.

Hooke believed that the British Isles may well have been raised out of the sea in the very same quake, exposing the large banks of ancient fossilised shellfish that were well known in the area. He also made use of more recent evidence to back up his theories, such as the effects of the Antilles earthquake of 1690, and the discoveries of elephant and hippopotamus bones in England, Pomerania, and Siberia. Unfortunately, these arguments failed to convince his contemporaries at the Royal Society, none of whom was willing to concede that earthquakes could have caused such drastic alterations to the Earth's topography. They were even less willing to accept that fossils were remnants of living organisms, and instead continued to invoke the *vis plastica* theory. For them to have believed otherwise would have meant that species were not permanent but were capable of becoming extinct, an idea that was far from popular at that time.

One of Hooke's academic associates was a Danish naturalist and physician, Niels Stenson (1638-1686). At the age of twenty- eight,

Stenson had been invited to work in Florence by the Duke of Tuscany, Ferdinand II de Medici. Even at this age, Stenson had already acquired a reputation as a brilliant scientist, having made intensive studies into the anatomy of the human brain and heart. Not long after he arrived in the city, he converted to Catholicism and adopted the name of Nicolaus Steno. He was employed as personal physician to the Duke, and teacher to his sons, but he still found time for one of his most satisfying pastimes, fossil collecting. One particular kind of fossil referred to as "tongue stones" were commonly found in Tuscany. Steno had seen many "tongue stones" and he noticed the resemblance between them and the teeth of modern sharks.

Fortuitously, a shark was washed up on a nearby beach in 1667, and Steno immediately seized upon the opportunity of studying its dentition. Afterwards, he was left in no doubt that the tongue stones were derived from ancient sharks. The details of the "tongue stone" teeth did differ from those of the modern shark however, and he decided that the species to which the fossil teeth had belonged was now extinct. In addition, the fact that the fossils were found on dry land meant that the area must have been submerged when the teeth were petrified. These conclusions convinced him to undertake a detailed study of the geological processes that cause sedimentation and fossilisation.

In 1669, after two years of research, he completed his book (in Latin), entitled *Prodromus to a Dissertation concerning a Solid naturally contained within a Solid*. The work dealt with many aspects of palaeontology and geology, and was almost immediately translated into English by the Royal Society. Some basic laws of stratigraphy emerged in Steno's writings, the major one being that if layers of sediment are deposited and remain undisturbed, the oldest ones will lie the deepest, while the youngest ones will lie closest to the surface.

Nicolaus Steno (1638-1686).

Thus the geological time sequence could be read from the base upwards. He also realised that deposits would always be laid down horizontally, and that deviation from this meant that the beds must have been altered from their original positions. Such occurrences could take place during earthquakes, rift movements, or other natural processes. Most important of all, however, was his observation that the fossils contained in the lower rock strata differed considerably from those contained in the strata above. They therefore represented animals (or plants) that no longer existed by the time the later layers were deposited.

Not long after completing his book, Steno joined the priesthood and began to neglect his work on geology and fossil study. After serving at several religious positions around Germany, he was transferred to the town of Schwerin where in the unsanitary conditions, he soon fell ill and died. Strangely enough, like Robert Hooke and Leonardo da Vinci so many years earlier, Steno's pioneering work went largely unnoticed by the scientists of his day. Many years would elapse before the true value of his work was recognised.

One of the most vocal and prominent supporters of the Deluge theory actually ended up being partly responsible for its eventual downfall. That person was a Swiss doctor from Zurich, Johann Jacob Scheuchzer. As a young student, Scheuchzer was, like many of his contemporaries, a dedicated believer in *vis plastica*, having had accumulated a modest collection of curious "figure stones". But after reading a book by John Woodward of the Royal Society, he began to have doubts. Woodward's book, *Essay toward a Natural History of the Earth* essentially tried to show how all tilted and displaced strata could be accounted for by a single world wide flood. Likewise, the fossils of marine and non-marine animals might all have been examples of the victims of that same flood, remains of animals caught up in a sudden rush of waters and deposited at random where they sank once the waters settled. Scheuchzer was suitably impressed by such ideas, and set to work translating the book into Latin. At the same time he began reorganising his own fossil collection to fit in with the concepts expressed by Woodward.

He then proceeded to write and publish a pamphlet entitled *Complaints and Justifications of the Fishes* in which he tried to discredit the *vis plastica* theory, a theory that he himself had previously subscribed to. Scheuchzer's collection of figure stones included some large vertebrae that he had collected some years earlier, and he included a drawing of them in the pamphlet, labelling it as the remains of a human victim of the Deluge. One of his former student friends totally disagreed with him and claimed that the bones were nothing more than fish vertebrae, but Scheuchzer would not be swayed. The bones, in fact, belonged to neither fish nor man, but originated from an ichthyosaur.

Somewhat later, in 1715, Scheuchzer came into possession of a large slab of rock containing part of an articulated skeleton within it, a skeleton that was destined to finally discredit the concept of the Deluge. The stone had come from a limestone quarry and was given to Scheuchzer by monks from a nearby monastery. Scheuchzer proudly proclaimed his new discovery as the skeleton of a man drowned in the Deluge, a rare relic of that accursed race of people of the "first" world. He named it *Homo diluvii testis*, the man who was witness to the flood. When several similar skeletons subsequently turned up in the quarries, people began to doubt that they were human remains. It was not until over 100 years later that the true identity of the "witness to the flood" became known. Upon viewing the skeleton, the notable French palaeontologist and anatomist Baron Georges Cuvier (1769-1832) immediately recognised it as belonging to no less an animal than a giant salamander from the recent Cenozoic Era, an animal that still survives in parts of Eastern Asia and now bears the name *Andrias scheuchzeri* (andrias meaning "man image" in Greek).

After Cuvier's refutation, many scientists openly questioned the authenticity of the Deluge theory, insisting instead that palaeontological research be carried out totally independent of theological dogmas. In future years, it was to be the evolutionary theories of the naturalists Jean Baptiste de Lamarck (1744-1829) and Charles Darwin (1809-1882) that were to have the greatest impact on palaeontology and biology. Lamarck's "Philosophie Zoologique", published in 1809, was the first substantial challenge to the creationists' view on life. In it, he advanced views that species were not immutable, but had changed and evolved through time, giving rise to other species in the process (while the exact mechanisms that govern the process of evolution have yet to be elucidated, the fact that new species have arisen throughout the geological record is indisputable). But like so many other notable figures in the history of palaeontology, the importance of Lamarck's work was never adequately recognised during

Jean Baptiste de Lamarck (1744-1829).

Charles Darwin (1809-1882).

his lifetime. He would die destitute and blind, and even his own countryman, Cuvier, (who for all his brilliance was a creationist who saw little value in Lamarck's theories), paid him a harshly worded tribute that had to be toned down before it could be issued by the French Academy of Sciences.

But after publication of Darwin's "Origin of Species by Natural Selection" in 1859, the creationists' concept that species were immutable very quickly became obsolete. Palaeontologists now realised that they should be able to study the fossil record by analysing relationships between fossils from different strata levels, rather than viewing each fossil in isolation. By using proven methods of study, they could carry out palaeontological research within the framework of sound geological and biological principles rather than under the constraint of a predetermined religious dictum.

2.3. The Ages Beneath our Feet

Fossilisation of most types of animals and plants is an exceptionally rare event, and the chance that any given animal will be preserved is only one in many thousands. Of those that do fossilise, there is, again, only a very small chance that they will ever be found by anybody, let alone someone who realises their significance. The fossil record is very incomplete and only provides a glimpse into the total number and diversity of extinct forms of life. Yet in spite of this, the very wide range of species represented in fossil collections recovered from all continents gives some concept of how various lineages of animals and plants have changed through time.

In rocks older than the Cambrian period, the most commonly encountered fossils are the stromatolites. These consist of small, branching structures formed of silica or calcium carbonate and are sometimes compacted into limestone beds many metres in thickness. Stromatolites are still being produced today in the world's warm, shallow oceans by the action of blue-green algae, a group of very basic, photosynthetic organisms. So it is reasonable to assume that plants of a similar nature also existed in the pre-Cambrian oceans.

The earliest forms of animal life were soft-bodied invertebrates that did not possess external shells, and the only evidence now remaining of them consists of impressions made in soft sands or muds that have since petrified. From these traces, it appears that they basically belonged to two phyla, the Coelenterata, which includes hydras, corals, jellyfish and sea anemones, and the Annelida or segmented worms. The age of these is pre-Cambrian, that is greater than 600 million years, and because the oldest rocks that have been analysed by radioactive dating techniques register ages of between 4.5 and five billion years, it is only the last eighth or so of the Earth's strata that contains evidence of animal life (though fossil bacteria and blue-green algae may well go back as far as 3.5 billion years).

In Lower and mid-Cambrian strata, the rocks contain many fossils of invertebrates most of which have mineralised skeletons of some sort. These include the Sarcodina (amoebas, foraminiferans, heliozoans, radiolarians), the Porifera (sponges), the Brachiopoda (lamp-shells), and the Arthropoda. The arthropods include a vast collection of forms such as the arachnids (including scorpions, spiders, harvestmen, ticks and mites), crustaceans, insects, millipedes, centipedes, and horse-shoe crabs. By far the most prolific of the Cambrian arthropods were the trilobites. These strange creatures resembled no living types, but in appearance can be superficially compared to giant wood lice with hard exo-skeletons and sophisticated compound eyes. It has been estimated that trilobites account for some 60 percent of all Cambrian fossils. Even after that period, they continued to thrive, leaving a most prolific fossil record along the way, until the Permian period, a total span of some 350 million years.

Moving up into the Ordovician strata, the rocks become characterised by an increase in the number of fossils left by shell-bearing invertebrates such as bivalves, gastropods, and crustaceans. Many of these animals developed shells as protection against the predations of another invertebrate group that increased in numbers at the same time, the Cephalopoda which include the octopuses, squids, and cuttlefish. From the Ordovician onwards through the rest of the Palaeozoic then on through the Mesozoic and finally into the recent Cenozoic, remains of the various vertebrate groups are found in increasing numbers and diversity. In chronological order of evolution and development, these are firstly the fishes, then the amphibians, the reptiles and mammals, and finally the birds. All of the vertebrates possess hard, osseous skeletons capable of long-term fossilisation.

The chance that a vertebrate might be preserved was increased greatly if it dwelt in or around habitats such as lakes, swamps, or coastal lagoons, because these areas usually possess the required conditions for the process of fossilisation. After death, the carcasses sometimes became trapped in muddy sediments being deposited by the movement of water. The gradual accumulation of silt or sand would build up over the carcass and entomb it. Soft parts of the body such as the muscles and tissue would have decomposed quickly, but under the action of different minerals in the water, the osseous remains hardened and even petrified, that is, changed in chemical composition. In the process of petrification, the bone is gradually replaced by new materials, such as lime or silica, but its shape and morphology remain unchanged. Sometimes, conditions in the water cause the bones themselves to dissolve, leaving hollow impressions in the sediment. Such impressions may themselves become fossilised.

The majority of fossils found consist only of fragmentary remains of an animal, such as a skull, or isolated teeth or vertebrae, and it is an exceptional event when a more-or-less complete skeleton is uncovered. The reason for this is that the body rarely remains intact after death. In a river situation, for example, the projecting parts of the carcass such as the skull, limbs and tail, often become separated from the torso and are washed away. Or if the carcass floats on the surface of a lake or swamp, the actions of scavengers and adverse weather conditions can speed up the process of decomposition, causing parts of the body to be lost before being covered by sediment.

Another type of fossilisation can occur in drier, semi-arid regions. The carcasses of animals that have died in such environments can dehydrate and become mummified. At some later stage, perhaps when the area is inundated by floods, the mummy can be silted over and though the skin itself might decompose, the impressions that are left in the silt might remain and harden. Some well-preserved skin impressions have been discovered, including one example of the remains of the hadrosaur *Anatosaurus*, uncovered in Wyoming in 1908.

As well as actual body remnants, any preserved traces or tracks made by animals during their lifetime also come under the description of a fossil. The remains of partly digested foods, tooth and claw marks, preserved dung (coprolites), and preserved eggs are some examples. Footprint tracks are sometimes uncovered, made perhaps in the still soft ash fall of a volcanic eruption, or on the muddy shore of some ancient swamp or lake. A series of tracks found in Lower Cretaceous rocks in Texas records a prehistoric life and death drama. It shows the prints made by a giant sauropod dinosaur whilst just alongside are the three-toed footprints of a bipedal carnivorous dinosaur. The relative positioning of the two sets of prints indicates that the carnivorous dinosaur was in fact following or perhaps stalking the larger sauropod, at times even stepping into its tracks (unfortunately, the outcome of this particular pursuit remains unknown as the trail leads eventually under a solid face of limestone and is lost).

Other clues might also be found within the same deposits that contain animal fossils. Soft plant and pollen remains, fossilised tree trunks, rain- and hail-pitted surfaces, or water ripple marks left on an old, now long-vanished coastline can all provide valuable information as to the types of vegetation, climate, weather patterns and general environmental conditions that prevailed at the time the animal lived.

Fossilised dinosaur eggs, particularly those from the Gobi Desert, have been found more-or-less intact as they were laid over 75 million years ago. Some even contain the remains of unhatched embryos. Finds such as these can help us to understand the habits of these reptiles as well as confirming that some dinosaurs were definitely oviparous. But unless the remains of the animal thought to be responsible for a trace fossil can be found in the same location and strata, and a close association between the two is evident, it is difficult to identify which animal was responsible for the trace. There is little doubt, for instance, that the numerous nests of eggs found in the Gobi were laid by the small dinosaur *Protoceratops*, because they were found together with newly hatched and juvenile members of this genus (see chapter three). Sometimes, because such evidence is lacking, incorrect conclusions can be drawn.

From the years 1835 to 1863, a New England professor, Edward Hitchcock, pondered over, studied, and described numerous dinosaur footprints that were scattered over the Connecticut Valley area. Locals referred to the curious tracks as the "footprints of Noah's raven". Hitchcock culminated his lifetime study with the publication of a monumental work, in which he proposed many new species of gigantic, extinct birds to account for all the different types of three-toed prints. No animal remains were ever found from the same area,

Some of the forty-nine different types of footprints exposed and described by Edward Hitchcock over a period of three decades. Thousands of similar prints are now known from northern Australia and, like those of the Connecticut Valley, are the distinctive tracks of dinosaurs.

so the professor never doubted their supposed avian origin.

The fossil record can be misleading in some other respects as well, principally by giving a biased view of the comparative numbers of different species from a given area. Species inhabiting low lying floodplains close to water will be relatively well represented in the fossil record as these will have comparatively high rates of preservation. Conversely, species from rolling steppe or hilly and mountainous areas will be only poorly represented, because elevated lands tend to suffer more from the effects of erosion due to wind and rain in addition to not usually having the conditions required for adequate preservation.

In the Pleistocene asphalt pits of La-Brea, the sabre-toothed tiger Smilodon *often became mired along with its intended prey, the elephant* Archidiscodon. *(Based on Z. Burian)*

Skeletal fossils are not the only body remains that can withstand long durations of time, though they are usually the only ones capable of lasting to the order of millions of years, because their altered chemical states are extremely resistant. In many areas of the Northern Hemisphere, there occur geological deposits that permit preservation of entire animals. Swamps, deposits of mineral oils or waxes, and asphalt lakes act to mummify the bodies of animals unfortunate enough to have fallen into them. At the Californian locality of Rancho-la-Brea, a Pleistocene lake rimmed with molten asphalt has trapped and preserved elephants of the genus *Archidiscodon*, as well as young, inexperienced sabre-toothed cats (*Smilodon californicus*) that had come to feed upon the mired elephants. From Europe, well known examples of such preservations include two woolly rhinoceros calves and a young mammoth from the recent Ice Ages that were unearthed in Galicia (Poland), all three animals having become trapped in a deep ozocerite (mineral paraffin) deposit.

Finally, some areas such as northeastern Siberia are subject to extreme climatic conditions which, together with the low temperatures, allow animals to become frozen after death, remaining in this condition for thousands of years. Many excellent cadavers of the Siberian mammoth (*Mammuthus primigenius*) have been unearthed in recent centuries, including the fully grown "Berezovka" mammoth found in the eastern Yakutsk area at the turn of the 20th century. The giant animal had apparently fallen into a crevasse and because of its size, was unable to extricate itself. In some cases, the freezing process has been so complete that the flesh remains quite edible, and participants at a geological congress at Petrograd (Leningrad) were once served with "fresh" mammoth steaks that had been cut from a young animal found frozen into a large block of ice and transported, as such, to the capital.

This last means of preservation may represent a chance of being able to recreate an extinct form of life. The tissue cells of some mammoth finds are so well preserved that scientists at the Petrograd Institute of Cytology believe that it may eventually be possible to obtain a culture of living mammoth cells. One of these cells could be fused with a sex cell from a female Indian elephant, the nucleus of which will have previously been destroyed by carefully focused X-rays. This new cell would then be implanted into a female elephant's uterus, hopefully resulting, some 18 to 20 months later, in the birth of a mammoth calf. To date, however, none of the frozen cells from mammoth carcasses have proven revivable.

The Siberian mammoth Mammuthus primigenius *from the recent Pleistocene. (Painting by the author, based on Z. Burian)*

2.4. Dinosaurs and Men

Although the first dinosaur remains were described in the 1820s by English naturalists, most work involving finding and describing new species was undertaken in the western United States, particularly in the states of Colorado and Wyoming between the years 1877 and 1900. The leading figures in this early pioneering era of dinosaur palaeontology were two energetic and headstrong rivals, Edward Drinker Cope (1840-1897), a brilliant anatomist, ichthyologist, herpetologist and palaeontologist from Philadelphia, and Othniel Charles Marsh (1831-1899), the first director of the Yale Peabody Museum. With each trying to outdo the other, they were responsible for bringing to light many new types of mainly Jurassic and also some Cretaceous dinosaurs.

Their work not only laid the broad foundations of dinosaur taxonomy, but also firmly established the concept of these great reptiles in the minds of the general public. The impressive list of genera found during what is often referred to as Marsh's and Cope's "bone war" included *Allosaurus, Ceratosaurus, Apatosaurus* (= *Brontosaurus*), *Diplodocus, Camarasaurus, Trachodon, Camptosaurus, Monoclonius, Triceratops, Stegosaurus, Nodosaurus*, and *Ornithomimus*. The final tally was nineteen genera for Marsh and nine for Cope, though this is a misleading comparison because Marsh had a full complement of museum staff at his disposal whereas Cope worked largely on his own. Certainly in the wider field of vertebrate

Skeleton of the late Jurassic plated dinosaur Stegosaurus, *one of many North American dinosaurs excavated by E.D. Cope and O.C. Marsh. (Photo by the author)*

palaeontology as a whole, Cope's achievements are unequalled, and he was the author of no fewer than 1,400 individual papers, books, and monographs! For palaeontologists in the United States, he is regarded as one of that country's founders of the science.

In 1878, a quite accidental but significant dinosaur find was made on the other side of the Atlantic. At a depth of some 300 metres down a Belgian coal-mine near Bernissart was found a collection of more than thirty skeletons of the Cretaceous ornithopod *Iguanodon*, the animals having fallen and become trapped in a ravine. By the early 1900s, palaeontological work in North America was continuing in the western states of Wyoming and Utah, as well as in the Badlands of the Red Deer River area of Alberta in Canada.

Also by this time, major late Jurassic dinosaur finds were being made in Tendaguru, East Africa (now Tanzania), by German workers under the leadership of palaeontologists Werner Janensch and Edwin Hennig. A series of large-scale operations during the period 1907-1912 included at its height over 500 native labourers and their families. Interesting new genera came to light, some of which exhibited close affinities to those excavated by Marsh and Cope, whilst being generally smaller in size (such as the stegosaur *Kentrosaurus*, and the sauropod *Dicraeosaurus*). However, the most notable find of the German excavations was a sauropod of unusually large proportions; *Brachiosaurus* (= arm reptile) was a daunting thirteen metres in height and some 22.5 metres in length. Along with much of the other Tendaguru material, this great skeleton of the largest completely-known dinosaur was mounted in the von Humbolt Natural History Museum in Berlin.

By the turn of the 20th century, interest in vertebrate palaeontology had also spread to Asia. Records of "dragon bones" from Sichuan in China actually date as far back as 265 A.D., though the formal study of dinosaurs in this area began in 1902 with collections made in northern China by Colonel Manakin of the Russian army. Remains of the hadrosaur *Mandschurosaurus* (now housed in Petrograd) were excavated from there by Russian workers between 1915 and 1917. Since 1933 however, palaeontological exploration of China has mainly been undertaken by Chinese scientists under the supervision of Yang Zhong-jian (known in the West as C.C. Young). Notable Chinese dinosaurs included the prosauropod *Lufengosaurus*, the hadrosaurs *Tsintaosaurus* and the gigantic *Shantungosaurus*, the incredibly long-necked

The enormous skeleton of Brachiosaurus *from Tendaguru, as reconstructed by Werner Janensch. A human skeleton stands beneath. (By permission of H. Jaeger, Von Humboldt Museum of Natural History)*

sauropod *Mamenchisaurus*, and the stegosaur *Tuojiangosaurus*.

Following the First World War, the American Museum of Natural History in New York, which had previously been involved in working the Upper Jurassic strata of Marsh's old stamping grounds in Wyoming, organised a series of elaborate and labour-intensive expeditions to the Gobi Desert of Mongolia and the Chinese territory of Inner Mongolia. Lead by Roy Chapman Andrews, these pioneering expeditions took place between 1922 and 1930. Although their original objective was to locate evidence of the origins of early humans and other mammals, they are today chiefly remembered for the discovery of large numbers of the small dinosaur *Protoceratops* and associated nests of eggs at Bayn Dzak. Other dinosaurs found included *Saurornithoides, Velociraptor, Oviraptor,* and *Psittacosaurus.* Some very early placental mammal remains were also uncovered, while Cenozoic mammals were collected from central Mongolia and Inner Mongolia. This material is housed at the American Museum of Natural History.

Immediately after the Second World War, the Gobi was again the venue for large-scale systematic expeditions, this time by Russian expeditions under the auspices of the Soviet Academy of Sciences. Between 1946 and 1949, a series of three major expeditions under the leadership of Ivan Efremov and Anatole Rozhdestvensky explored large parts of Mongolia and uncovered many fossil localities, the most important being the unusually productive Nemegt Valley. The strata of this area are deltaic deposits from areas inhabited by large numbers of dinosaurs during the late Cretaceous. Complete skeletons of giant dinosaurs found included the hadrosaur *Saurolophus* and the tyrannosaur *Tarbosaurus,* as well as many new types of ankylosaurs, theropods and other dinosaurs. In addition, large collections of mainly Cenozoic mammals were excavated from western and central Mongolia. Material from these expeditions is housed both at the Russian Academy of Sciences' Palaeontological Museum in Moscow and at the Museum of the Mongolian Geological Institute in Ulan Bator.

In the 1960s this work was followed by a series of joint Polish-Mongolian expeditions under the leadership of Zofia Kielan-Jaworowska, proceeding in three periods of field work (1963-1965, 1967-1969, 1970-1971). Work was conducted at Nemegt, Bayn Dzak and at other localities, and as well as the collection of well-known dinosaurs, there were also many totally new finds. These included

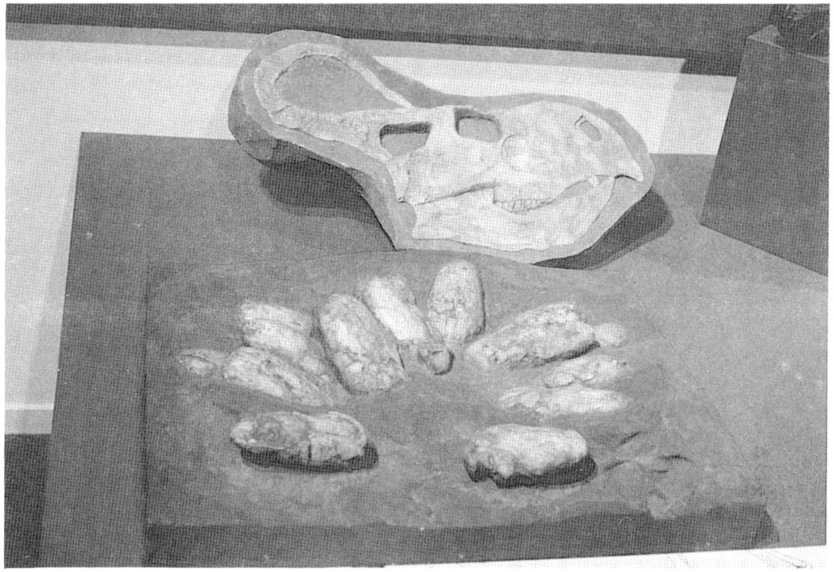

Reconstructed nest of the late Cretaceous Mongolian dinosaur Protoceratops andrewsi, *and cast of the skull from an adult. (Photo by the author)*

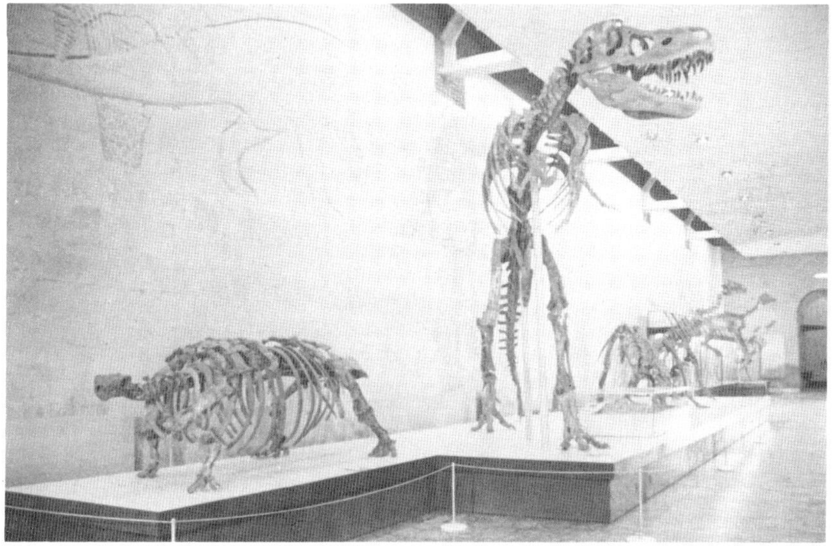

Mongolian dinosaurs mounted at the Moscow Palaeontological Museum. (Courtesy of S.M. Kurzanov, Russian Palaeontological Institute)

ornithomimosaurs, pachycephalosaurs, sauropods, and various theropods. Collections of both Cretaceous and Cenozoic mammals were also made. Material from these expeditions is housed at the Polish Academy of Sciences' Palaeobiology Museum in Warsaw and at the Museum of the Mongolian Geological Institute.

The Gobi Desert has proven to be an unusually prolific area for dinosaur remains, particularly with respect to theropods, and finds made here have resulted in the creation of a new infraorder (the Segnosauria, Barsbold and Perle 1980), a new suborder (the Pachycephalosauria, Maryanska and Osmolska 1974), several new families, including Deinocheiridae (Osmolska and Roniewicz 1969) and Saurornithoididae (Barsbold 1974), and of course many new genera and species. What follows then is a narrative of dinosaur collecting in the Gobi Desert, what the American palaeontologist Edwin Colbert has called "the Asiatic dinosaur rush". It is divided into three chronological chapters, 1) The American Museum of Natural History Expeditions, 2) The Russian Expeditions, and finally 3) The joint Polish-Mongolian Expeditions. This last chapter also includes some of the recent results from more protracted field work, such as the joint Russian-Mongolian Expedition series.

Chapter Three

To the Ends of the Earth

3.1. On the Trail of Ancient Man

BY THE late 19th century, the science of vertebrate palaeontology had developed to a reasonably advanced level both in the United States and in some European countries. One of the main topics of debate to scientists on both sides of the Atlantic was the evolution of the placental mammals, the group that contains the majority of mammal species. Modern mammals can be classified into three main groups, the Prototheria (monotremes), the Metatheria (marsupials), and the Eutheria (placentals), all of which are believed to have originally evolved from the mammal-like therapsid reptiles. The monotremes are the least evolved mammals, and like their reptilian ancestors, they are egg-laying animals. Surviving examples are the web-footed aquatic platypus and the spiny anteater or *Echidna*, both from the Australian region. The more advanced marsupials are also represented in that part of the world in forms such as the kangaroo, wombat, wallaby and the koala "bear", but they are also found in America, in a variety of opossums and related types. In contrast to the egg-laying monotremes, marsupials give birth to live young at a very tiny and helpless stage of development. In most cases, they then have to be carried for some time within a pouch on the outside of the mother's abdomen.

All other mammals belong to the Eutheria which has been by far the most successful group. The success of the eutherians is due to their ability to give birth to young which are relatively large and well developed. This feature was made possible due to the allantoic membrane (which had previously surrounded the developing eggs of their reptilian ancestors) having come into contact with the uterus containing the developing foetus. Through this, the young mammal is able to receive supplies of oxygen and nourishment from its mother during a long gestation period, thus allowing it to grow considerably within the security of the uterus. The name "placental" is an ill-chosen term for the eutherians, because marsupials also possess a placenta of sorts, though the gestation period that it permits is far shorter than that of the eutherians.

19th century scientists knew that the Paleocene strata of North American and European deposits had yielded a rich placental fossil fauna, but that the previous Cretaceous strata during the Age of Reptiles appeared to be devoid of them. However, the development already evident in many of the Paleocene placentals suggested that their origin must have lain further back well within the Mesozoic. A general consensus was that the first placentals had evolved on one particular continent, perhaps during the Cretaceous, then later migrated to the North American and European landmasses in the early Paleocene.

At the turn of the 20th century, Prof. Henry Fairfield Osborn, eminent palaeontologist at the American Museum of Natural History in New York, advanced the suggestion that the place to search for this pre-Cenozoic placental development should be Central Asia, an area that was then completely unknown territory with respect to palaeontology and most other sciences for that matter. A decade later, Dr. William Diller Matthew of the same institute took up this view again in his monumental work, "Climate and Evolution". Based upon his detailed knowledge of the distribution and taxonomic relationships of both extinct and extant animals, Matthew envisaged Central Asia as a kind of Mesozoic Garden of Eden for the earliest placentals, a place where they originated, evolved, and eventually migrated from.

One of the American Museum's zoologists at that time was Roy Chapman Andrews. He was so inspired by Osborn's and Matthew's arguments that he conceived and promoted the idea of a series of elaborate expeditions to Central Asia, with the primary goal being to throw light upon the origins and evolution of the earliest hominids. Since 1912, Andrews had been involved in zoological exploration in different corners of Asia, and Osborn's prophesies had always influenced his work. In 1915, he had presented a proposal to the president of the American Museum for a ten year series of zoological collecting expeditions to Central Asia, and the first of these, the 1916-17 expedition to southwest China and the Tibetan border, had yielded an impressive collection for the Museum. Following World War I, Andrews spent part of the 1919 summer season within Mongolia on the second expedition, and whilst there, was constantly impressed by the similarities of the Asiatic fauna to those of Europe and America. Such similarities always appeared to bear out Osborn's and Matthew's theories regarding the origins of the early mammals.

Andrews realised that if a series of archaeological and palaeontological expeditions were to be successful in Central Asia, they must include not only specialists from those disciplines, but from other related sciences as well. The geology, botany, topography and zoology of the area would all be of relevance if a complete picture of prehistoric life was to be elucidated. Not all of Andrew's colleagues were enthusiastic toward his ideas however, because the existence of any fossil-bearing strata at all in this part of the world had yet to be proved. Many forecast that Central Asia would turn out to be a vast, unfossiliferous sandy waste of gravel and thorn bush, consuming time and expense in large amounts. Andrews was warned that if the searing heat of summer did not prove an obstacle to effective field work, then the consequences of being trapped in the Gobi during the freezing winds of winter would be even more dire.

Although several naturalists and geologists had previously ventured into the region and crossed sectors of the Gobi Desert, there had been but one report of fossils in the entire area. In 1892, the Russian geologist and explorer Vladimir A. Obruchev had found some rhinoceros teeth on the old caravan route that ran from the Chinese border to the Mongolian capital, Urga. But by the 1920s, another Russian, the palaeontologist A.A. Borissyak, had located a rich deposit of Cenozoic mammalian fossils in nearby Kazakhstan. Correlation of the geology of Kazakhstan with that of the Gobi indicated that the desert could well yield some interesting Cenozoic and possibly Cretaceous fossils as well.

Andrews calculated that five years and a sum of $250,000 would be the minimum framework from which to begin any serious planning. In spite of the enthusiastic support shown by Osborn when Andrews

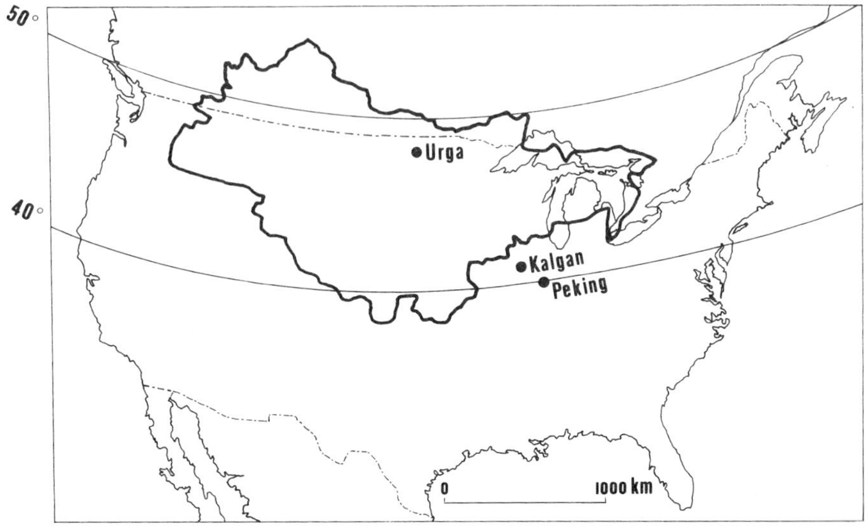

The great distances involved when conducting explorations within Mongolia are evident in a map overlay comparing the country with the United States.

presented his plans in 1920, it was evident that the American Museum could not afford to finance the project. Most of the backing would have to be obtained from private individuals prepared to risk their money on such a venture, and Andrews had to be willing to "hard sell" the Central Asiatic Expeditions. He later wrote that by February 1921 he had already spoken at so many dinners and luncheons, interviewed so many financiers and written and lectured so much in promoting the expeditions that they had already taken ten years off his life before even starting. On top of this, the sheer logistics of working successfully with a large group of people in the middle of the Central Asian desert was to prove even more of a daunting prospect, requiring organisational approaches never before attempted on such a scale.

Mongolia was a largely unknown and totally undeveloped country half the size of Western Europe. Lying isolated within the heartland of Asia, even by 1920 the country did not have a single mile of railway within its borders. Because of the harsh winter conditions, effective field work could only be carried out between the months of April and October. Although past explorers had always used camels for transport, averaging some sixteen kilometres per day, Andrews believed that sturdy automobiles could also be used in desert conditions. These could cover as much as 160 kilometres per day, moving into the desert as soon as the winter snows had departed, and moving back out again before the temperatures dropped the following season. After considering several European makes, Andrews chose Dodge Bros. cars of Detroit (fitted with extra heavy-duty springs), while Fulton one-ton trucks (New York) would be used for any heavy hauling. Not surprisingly, no insurance company would provide cover for any of the vehicles, but the fleet of cars would eventually end up covering over 1000 kilometres of mostly unknown terrain, and were later sold to a Chinese company and continued in service for some time.

Because the cars would be constantly required for transport and observation work, they could not afford to be tied up hauling supplies as well, so the time-tested method of camel transport could not be dispensed with. The plan was to send large camel caravans across the desert along predetermined routes, hauling all sorts of food and supplies including many five-gallon cans of petrol and oil. The scientific staff would then use the cars to traverse the desert in zig-zag paths in small groups, searching and collecting all the while. With the absence of distinguishable landmarks, directions and locations were to be ascertained using the compass and the sextant, so navigation in this "sea of sand" was conducted exactly as it was on ocean-going vessels. At certain intervals, at predetermined points, the cars would rendezvous with the caravans, taking on food, petrol, and oil, and off-loading fossils and other collections onto the camels which were then led back to base camps. In one season, for example, 125 camels (each capable of transporting 180 kilograms) carried 17,600 litres of petrol, 450 litres of oil, three tonnes of flour, and 1.5 tonnes of rice, these being deposited at two depots, the furthest some 1,300 kilometres out from base.

3.2. The Desert Beyond the Wall

The headquarters for the American Expeditions was set up in the Chinese capital Peking (Beijing), and Andrews spent twelve months in preparation here. Apart from the vast assortment of scientific equipment required for effective field work in the desert, supplies included sturdy cotton tents and felt-lined Mongolian-type yurts, special heavy-duty coats, trousers and sleeping bags all sheepskin-lined, powdered eggs, dried milk, dried vegetables shipped from New York, and an array of canned foods and cured meat. Not least of all was a small armoury of guns required for hunting game, and should the need arise, to be used as protection against bandits.

By the spring of 1922, the first of the American Museum's Central Asiatic Expeditions had left Peking, bound for Mongolia. Apart from Andrews, who held the post of zoologist and leader, other staff members of the 1922 season included chief palaeontologist and second in command, Walter Granger (who had been a key figure in the

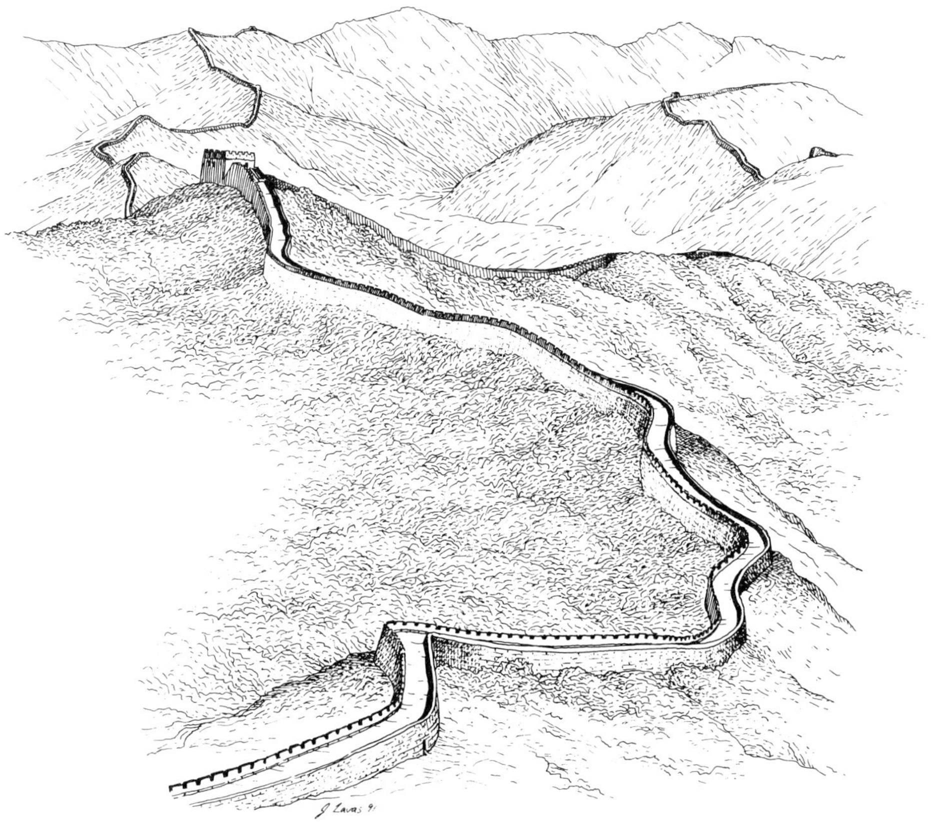

American Museum excavations for Jurassic dinosaurs in Wyoming between 1897 and 1902), chief geologist and petrographer, Charles P. Berkey, assistant geologist Frederick K. Morris, photographer J.B. Shackelford, chief of motor transport S. Bayard Colgate, and F.A. Larsen, interpreter. Also, the noted herpetologist Clifford H. Pope held the post of assistant zoologist for the Chinese division. In addition, there was a number of hired Chinese and Mongolian assistants that boosted the total number to twenty-six.

From Peking, the group travelled overland to the frontier town of Kalgan, commonly known as the "gateway" to Mongolia. On the morning of April 21, the expedition left Kalgan and made for the greatest of the Earth's man-made barriers. Originally built as a bulwark against invading Tartar tribes, from afar the Great Wall of China appeared like some giant prehistoric serpent, effortlessly embracing hills and mountains, and spanning valleys and rivers as it snaked its way over more than 2,200 kilometres of the Asian horizon. Passing beneath the narrow gateway in the wall, Andrews looked back upon the blue haze of the Chinese Shansi Mountains behind them, and at the sweeping expanse of undulating plains and gently sloping hills of Mongolia that stretched before them. Exactly what lay beyond was anybody's guess as the expedition convoy of three cars and two trucks set out in a northwesterly direction, bound for the Mongolian capital of Urga.

Ready for the journey into Mongolia, a car of the 1923 Central Asiatic Expedition waits at the northern gate of Kalgan, China. Henry Fairfield Osborn is seated at the front. (Courtesy department of library services, American Museum of Natural History)

In four days, the motor caravan had travelled 425 kilometres and reached the Iren Dabasu Basin, almost half-way between Kalgan and Urga. The lunar-like landscape of this area was even more desolate than the terrain they had been passing through. Conical mounds of wind-blasted sand arose from the ground like giant termite colony mounds, while stagnant salt marshes filled the lowest parts of the basin. Normally, Chinese workers extracted salt from here, hauling it in ox-cart caravans back to Kalgan, but these were not normal times. Mongols had been organising and arming themselves, and even though greatly outnumbered, had recently succeeded in ousting the Chinese army from their country. Throughout these expeditions, Andrews and his colleagues would come across remnants of vicious battles fought between Mongols and their former overlords. Torn Chinese uniforms, spent cartridges and scattered heaps of rifle shells were often all that remained of these unknown conflicts fought in the dead of night in an unknown land. Now, Iren Dabasu was found to be abandoned, the Buddhist temple and adobe huts of the local telegraph station deserted. However, the supply of petrol brought in by the camel caravan was awaiting them as planned, the caravan having passed through some two weeks before.

While the camp tents were being pitched, Berkey, Granger, and Morris drove off to explore some nearby outcrops. At sunset, they returned with Granger smiling broadly. Walking up to Andrews, he dug into his pockets, pulled out a handful of bone fragments and announced . . . "Well Roy, we've done it, the stuff is here!" A jubilant Andrews had been proved right and his critics proved wrong. Bones appeared to be present in reasonable quantities at Iren Dabasu. Notable were those of ancient rhinoceroses and titanotheres, the latter being rhinoceros-like horned mammals whose remains had previously been collected in the Eocene and Oligocene deposits of North America. This attested to the existence of the Bering land bridge between North America and eastern Central Asia at these times. But the unexpected sensation occurred the following morning when Berkey uncovered a large reptilian femur. Later, Granger uncovered a similar-sized tibia

Although some early titanotheres were no larger than sheep, the North American genus Brontotherium *exceeded a rhinoceros in size and lived in large herds upon the open plains during the late Oligocene.* (Based on Z. Burian)

from the same area. There was no doubt that the bones were those of a dinosaur.

Significantly, these were the first dinosaur remains and the first Cretaceous strata encountered in any part of Asia north of the Himalayan range, and in effect added an entirely new geological period to the then known strata of the area. But this being only a reconnaissance expedition, the group had to push on to Urga without delay. Granger and some of the others would stay on at Iren Dabasu for a week, while Andrews and the remaining expedition members would proceed to the capital where they would wait for Granger's group.

Upon reaching Urga, the expedition met with the Mongolian Minister of Justice, Mr. T. Badmajapoff, a quiet and charming personality who was to accompany the Americans from then on. Indeed, it is doubtful whether Andrews and his colleagues would have been able to reach many of the inner territories without the aid of Badmajapoff, as among other things, he was responsible for procuring the necessary permits required by his government for travel and field work within the country. A meeting was then arranged between Andrews and the newly installed Mongolian Premier and his cabinet (the country had rebelled against Buddhist lama rule and declared itself a Republic the previous year). During the meeting, the Premier asked Andrews if it was at all possible for the expedition to capture a specimen of a creature known as the "allergorhai-horhai". Although none of the cabinet members admitted to having seen one, it was solemnly described as a cylindrical-shaped animal about .7 metres in length and without discernible head or legs. Said to be so poisonous that to merely touch it was fatal, it inhabited the most desolate parts of the Gobi. Andrews soon realised that such an animal probably represented to the Mongols what the mythical dragon represents to the Chinese. Upon assuring the Premier that he had a pair of steel forceps suitable for seizing the allergorhai-horhai, whilst at the same time wearing heavily-tinted glasses to protect him from looking directly at such a potent creature, the meeting concluded with the best of feelings all round, and Andrews now knew that any possible bureaucratic barriers to the success of the expeditions had been avoided.

The Americans found Urga to be an unusual, rather bizarre town with an odd mixture of Russian, Chinese, and Mongolian architecture and customs. Spread over the north bank of the Tola River, this melting pot of some 60,000 inhabitants was a haphazard jumble of palisaded compounds surrounded by fluttering prayer-flags, felt-lined Mongol yurts, small cluttered Chinese shops, and ornate Russian houses. Amid the narrow and congested alleyways, Tibetan pilgrims and Manchu Tartars mingled with lamas, Turkestan camel drivers and Mongol tribesmen. However, one peculiar old Mongol custom added a darker side to life in some outlying areas of Urga. Mongols considered a human corpse to be the abode of evil spirits, and the immediate intention of the relatives was to dispose of the body as soon as possible. In the desert, this often entailed simply placing the corpse upon a horse-drawn cart and driving rapidly over rough ground so that it eventually fell off. The driver would hurry on without looking back lest he attracted the attention of the spirits within the corpse. Dogs, wolves and birds would quickly consume the body and leave the gleaming bones which every native would henceforth always shun. Some areas of Urga, particularly the base of the hill upon which the old lama city was built, were areas where corpses were abandoned, and the great black dogs that lived off the bodies became so

Participants on the second Central Asiatic Expedition of the American Museum of Natural History at Irdin Manha in 1923. Back row, third from left, Albert Johnson, third from right, George Olsen. Middle row, (left to right) Walter Granger, Henry Fairfield Osborn, Roy Chapman Andrews, Frederick K. Morris and Peter Kaisen. (Photo by W. Granger, courtesy of department of library services, American Museum of Natural History)

accustomed to the taste of human flesh that they were liable to attack any human upon the slightest provocation. Both Berkey and Andrews (together with Andrews' wife) were attacked on separate occasions, Berkey having to shoot two dogs with his revolver to save himself from being pulled down.

After leaving Urga, the motorcade turned toward the southwest, bound for the open desert. At certain points, zoological collections of some of the larger desert mammals and birds were undertaken by Andrews with the aid of poisoned bait. In June, one carcass liberally laced with strychnine poisoned no less than two wolves, four kites, a golden eagle, and a huge Gobi vulture. Members of this latter species, *Vultur monarchus*, can measure up to three metres in wing-spread and are particularly gluttonous scavengers. Andrews once observed one gorge itself so heavily on a dead ass that it could not become airborne, while on another occasion one consumed half of an antelope weighing 125 kilograms in less than thirty minutes! After many weeks of observation and collecting far out in the central Mongolian Gobi, the expedition turned back to begin the return journey to the headquarters at Peking. By that time, the Chinese capital was being wracked by political instability, having already been the venue of a major battle between two bitter rivals, Chang Tsolin and Wu Pei-fu.

In September, the party was driving eastwards looking for an old trail after not encountering any nomads for the last 160 kilometres. When three isolated yurts were sighted, Andrews went to make inquiries while Shackelford walked to the north for about a kilometre to investigate some curious mound-like structures visible on the horizon. He soon found himself looking over the edge of an extensive, red sandstone topographic depression that extended for some 18 kilometres in diameter. The edges of parts of the basin dropped away dramatically, and were totally hidden from the trail above. Half-way down one slope, Shackelford noted what appeared to be a bleached skull protruding from a sandstone pinnacle. When he had retrieved it and showed it to the palaeontologists, they realised that the twenty-centimetre-long skull belonged to an entirely new type of reptile, and so the two remaining hours of daylight were spent fossil hunting. There were many small isolated bones within the sandstone matrix of the basin, as well as a large number of eggshell fragments, which initially led Granger to believe that they had stumbled upon a Cenozoic deposit of birds' nest remains.

That evening, the expedition witnessed a brilliant spectacle above them. The heavily-weathered sandstone cliffs and flat-topped bluffs around the ramparts of the great basin seemed to capture and radiate the last rays of the setting sun, bathing the entire area in an awesome fire-red glow as if the rocks themselves were alight. The expedition members named this area the "Flaming Cliffs" of Shabarakh Usu (it is now officially known as Bayn Dzak). The colossal sandstone buttresses of the Flaming Cliffs were described by Andrews as resembling the spires and turrets of medieval castles, with huge gateways, walls and

ramparts. Everywhere, caverns ran deep into the rock faces which formed a labyrinth of ravines and precipitous gorges, a veritable haven for palaeontologists. One formation was named "the dinosaur" because it resembled a giant brontosaur slumped on its haunches. But the end of the field season was upon them, and with the cool autumn weather fast approaching, the group had to push on without delay, leaving any further work at the beautiful Flaming Cliffs for the following season.

Bearing supplies for the Central Asiatic Expeditions, a caravan of 125 Bactrian camels is dwarfed by monolithic sandstone buttresses at the foot of the "Flaming Cliffs" of Bayn Dzak. (Photo by J.B. Shackelford, courtesy of department of library services, American Museum of Natural History)

3.3. Where the Dinosaurs Hid Their Eggs

1923 saw the Central Asiatic Expeditions back in the desert for a further season of work, this time with an expanded team that included palaeontologists George Olsen, Peter Kaisen and Albert Johnson (who had experience in working the Red Deer River dinosaur-bearing deposits in Canada). Some weeks were spent excavating in the deposits of Iren Dabasu, where more isolated dinosaur remains were collected, but the expedition members were keen to return to the remote beauty of the Flaming Cliffs locality which they were forced to abandon so prematurely the previous season.

The basin was reached by July 8, and the Americans now knew exactly what to search for. After they had returned to Peking the previous season, the unusual skull found by Shackelford had been forwarded to the American Museum of Natural History. In due course, Osborn had cabled back to Andrews: "...You have made a very important discovery. The reptile is the long-sought ancestor of *Triceratops*. It has been named *Protoceratops andrewsi* in your honor. Go back and get more". The new find was described by Drs. W.K. Gregory and C.C. Mook, and was considered a possible ancestor to all the larger, horned ceratopians of North America (*Protoceratops* = first top horn-face). Skulls of *Protoceratops* appeared to be represented at the Flaming Cliffs in abundance, and by the time evening had arrived, almost every member had found "his own" skull.

On July 13, however, an even greater discovery was made, one that was to bring international recognition to these expeditions. Olsen reported to the others that he had found a group of what appeared to be three complete, large brown eggs fossilised into a sandstone matrix. It begged the question of whether these were birds' eggs after all. They appeared in the same strata as the *Protoceratops* remains but a primitive *Archaeopteryx*-sized bird could hardly be expected to have laid such large eggs. At first, Olsen thought that they might simply have been some type of geological artefact. But after Granger had examined the eggs, he could only arrive at one conclusion — "No dinosaur eggs have ever been found, but these must be dinosaur eggs. They can't be anything else". Palaeontologists had long assumed that dinosaurs, like many modern reptiles, were egg layers, but at that stage no such eggs had ever been identified. The possibility that these eggs might be dinosaur in origin plunged the expedition into an egg-hunting frenzy.

As digging proceeded, the basin revealed great numbers of fossil eggs, both fragmentary and complete, and even in groups arranged in apparent nest-like patterns. Andrews was surprised to find that in spite of the enormous significance of their find, they were evidently not the first discoverers of the eggs. One Mongol woman whose yurt was not far from the camp dutifully arrived daily with handfuls of eggshell fragments when she found they could be traded for empty tin cans, a scarce commodity in the Gobi Desert! In fact, remnants of Palaeolithic dwellings uncovered nearby showed that ancient peoples used these same dinosaur egg fragments cut to twelve-millimetre-square pieces as ornaments and jewellery such as necklaces. Andrews speculated that they were probably collected from the Bayn Dzak deposits.

Measuring about twenty centimetres in length and eighteen in circumference, the Bayn Dzak eggs were elongate and blunt-ended, shaped not unlike those of some modern lizards. The outside surfaces had a rough, almost corrugated or striated texture while fragments revealed a thickness of about 1.5 millimetres, and a smooth inner surface. Most of the eggs had small cracks in them, and these had permitted sand to filter inside and later solidify into sandstone. In two of the eggs, what appeared to be the remains of delicate bones could be seen. Although it has often been stated that these eggs contained unhatched embryos, later laboratory analysis showed them to be only artefacts of preservation. It was not until almost fifty years later that Sochava confirmed dinosaur embryonic remains within eggs recovered from the eastern Gobi during a Russian Expedition in 1971.

Many of the larger clusters of *Protoceratops* eggs revealed something of the nesting habits of the dinosaurs that laid them. One of the nests contained an inside circle of five eggs, surrounded by another circle of eleven, with two further eggs outside of this, indicating perhaps another third even larger circle. The eggs of some modern reptiles, such as those of marine turtles, are deposited in shallow depressions laboriously dug out by the females on isolated and naturally protected sandy beaches. After laying has taken place, the turtles refill the depressions and abandon the nests to return to the ocean, leaving the eggs to be incubated by the sun.

Female *Protoceratops* may have performed very similar nesting rituals along the shores of an ancient lake or river, over 75 million years ago. However, the eggs were not deposited at random as they are by turtles, but were either deliberately positioned or adjusted following laying. After digging a depression in a suitable sandy area, one female slowly deposited the eggs in a small circle, the wider ends of each egg tilting upwards. She may have continued to lay in a further circle outside of this, and it is possible that other females also deposited their eggs in the same nest, adding larger concentric circles of eggs to the initial one. This is thought to have been likely simply because some nests contained too many eggs for one animal to lay. Finally, the nests were concealed with sand by the mothers, probably by kicking backwards with their feet. As well as serving as an incubator, the sand also concealed the eggs from would-be egg-thieves such as small theropods. It is not known whether the adults abandoned their nests after concealing them, though as they appear to have lived in close proximity, their presence would have provided some protection.

At Bayn Dzak in the late Cretaceous, a communal nest containing the eggs of several Protoceratops *is covered with sand by a female and left to be incubated by the sun. (Based on J. Sibbick)*

We can surmise that the conditions required for the incubation and hatching of dinosaur eggs were probably very similar to those of modern reptiles. However our knowledge in this field is not great. It was previously believed that for incubation to be successful, the covering material should ideally be porous and loose to permit the ready conduction of air and heat. In the seasonal climates of today, some reptiles go to great lengths to ensure correct incubation conditions, but it is now known that in some cases, reptiles' eggs can actually be entombed in a fixed matrix and still be successfully hatched. For example, the Australian Lace Monitor (*Varanus varius*) employs the services of social insects when it lays its eggs within the hollows of termite colony mounds. The eggs are eventually enclosed within the sticky matrix constructed by the mound's inhabitants which ensures a warm, un-varying environment for the nine month incubation period (during which the temperature outside the mound drops markedly). When the young monitors are ready to hatch, the mother returns to the mound. After locating the completely concealed nest within, she uses her sharp talons to break the mound open, thus releasing the twenty-eight-centimetre-long hatchlings who are then left to fend for themselves.

Complete skeleton and skulls from various age groups of Protoceratops andrewsi *excavated from the Djadokhta Formation at the Toogreekiyn Shireh locality. This exhibit is in the Moscow Palaeontological Museum. (Courtesy of S.M. Kurzanov, Russian Palaeontological Institute)*

Olsen and Andrews (right) inspect a nest containing twelve Protoceratops *eggs during the 1925 season's field work at Bayn Dzak. (Photo by J.B. Shackelford, courtesy of department of library services, American Museum of Natural History)*

But in the largely non-seasonal conditions that prevailed for most of the Mesozoic, reptiles did not need to employ such involved nesting procedures. Like the marine turtle, small dinosaurs could have relied on shallow, sandy depressions dug on a secluded site. Not all sand would have been suitable however, that found within many of the eggs at Bayn Dzak being of an unusually uniform quality and grain size. *Protoceratops* was apparently a conscientious nest-builder and would have actively selected areas with sand of sufficient quality. Apart from this, there is geological evidence that during Cretaceous times there were small lakes to the south, where the eastern Altai Mountain ranges now stand. One or more of the draining rivers from here probably passed through Bayn Dzak, watering what would otherwise have been a somewhat dry inland locality (pieces of fossilised timber identified within the same strata appear similar to types of desert vegetation). Remains of small crocodiles and pond turtles have also been found in the Bayn Dzak vicinity.

Evidently, available herbaceous food, water and suitable sand all combined to make this area a favoured dinosaur nesting site in the Cretaceous. Other types of eggs present in the same strata show that *Protoceratops* was not the only reptile to use Bayn Dzak. One nest contained eggs twenty-three centimetres long and of a perfectly elliptical shape, with shells of three millimetres thickness. Eggs from another nest had shells of a very fine, almost paper-thin thickness while yet other varieties had smooth or pebbled and pitted surfaces. That Bayn Dzak was utilised season after season over a great period of time is verified by the separation of over sixty metres between the lowest and highest eggshell-containing strata.

But why did so many of the delicate eggs remain unhatched and preserved for so long? Andrews provided an explanation in terms of climate and geology. After the eggs were covered by the adults, freak storms or strong gales might have piled up around 1.5 metres of sand over the nests. As well as causing incubation to cease (the eggs now lying too low to benefit from the sun's warmth), this extra weight of sand would also have caused the shells to crack, thereby allowing the fluid contents to escape and the finer sand to enter. Another more likely explanation however, is that the increasing seasonality of the climate toward the end of the Cretaceous exposed an inadequacy in the nesting habits of herbivorous dinosaurs like *Protoceratops*. Like the eggs of the Australian Lace Monitor, some dinosaur eggs may have required long incubation periods. During most of the Mesozoic, this was not a problem, but the beginning of cooler autumn and winter seasons toward the end of the Cretaceous may have resulted in many clutches remaining unhatched, their tiny occupants having perished through heat loss during the cooler months. The innovative amniotic egg that had permitted the ascendancy of the reptiles in the Permian may well have become the tomb of embryonic dinosaurs in the late Cretaceous. Nests like those of *Protoceratops*, which were composed of eggs all laid in one horizontal layer, meant that a cold spell would have dealt equally harshly to all eggs. It is interesting to note that crocodiles and alligators which are archosaurs that have survived the Mesozoic, often lay their eggs in multi-layered tiers and/or carry out other procedures to ensure consistent heat exchange in their nests.

Whatever the reasons, the result was that a great many *Protoceratops* nests have remained almost exactly as they were laid during the late Cretaceous. Geographical changes and the associated development of a more arid climate eventually consolidated the whole sequence of strata into red sandstone with the eggs, both broken and whole, retained within the matrix. Without doubt, their discovery was the unexpected highlight of the American Museum Expeditions, and it is with these finds that the expeditions are always intimately associated. Today, dinosaur eggs are more common; subsequent expeditions to the Gobi have retrieved many more eggs of various dinosaurs, while nests are now known from other countries as well, including neighbouring China, from Provence in southern France (sauropods), and from Montana in the United States (hadrosaurs). But in the 1920s, the finding of a single identifiable dinosaur egg, let alone many nests of eggs was nothing short of a palaeontological sensation.

Of equal importance to the discovery of dinosaur eggs were the great number of excellently preserved *Protoceratops* skeletons found within the same matrix. With over 100 skulls and skeletons recovered, this quantity surpassed even the much celebrated group of *Iguanodons* found in the 19th century in Belgium. What made the finds even more significant was the range of growth stages represented, from newly-hatched young right through various age groups to fully-grown and old adults. For the first time, it was possible to study the changing body proportions of a species of dinosaur at many different growth stages. Adult *Protoceratops* were quadrupedal, heavily-built reptiles about two metres in length, with large heads and parrot-like "beaks".

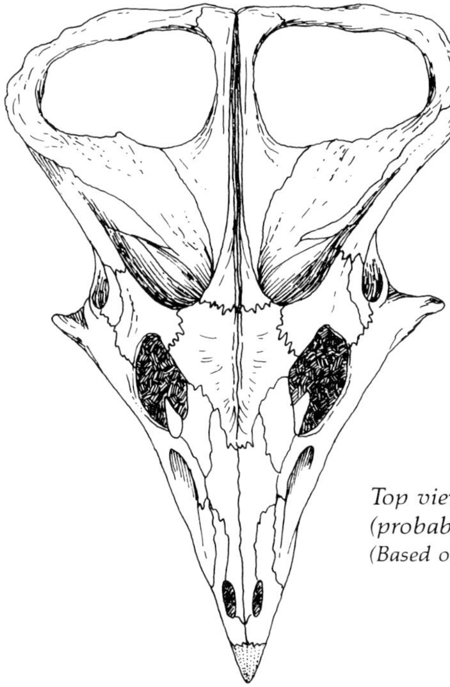

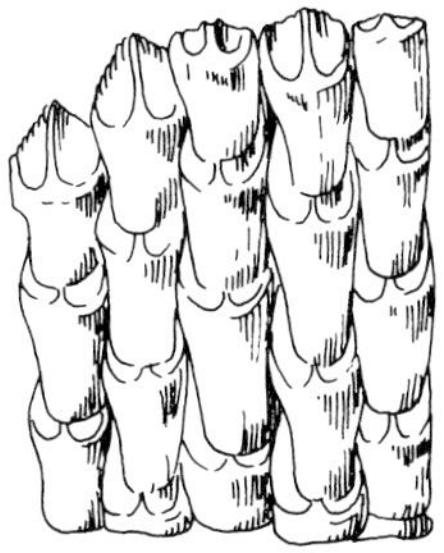

Layered tooth batteries of Protoceratops andrewsi. *(Based on Colbert)*

Top view of the skull of an adult (probably male) Protoceratops andrewsi. *(Based on Brown & Schlaijker)*

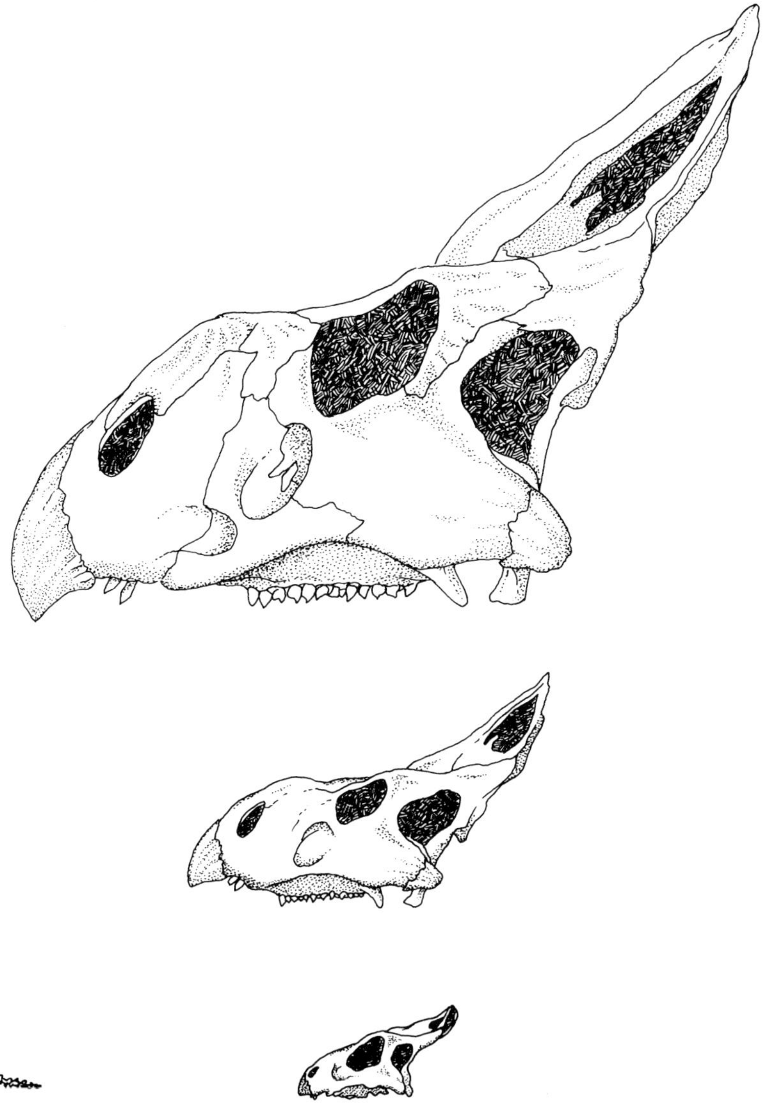

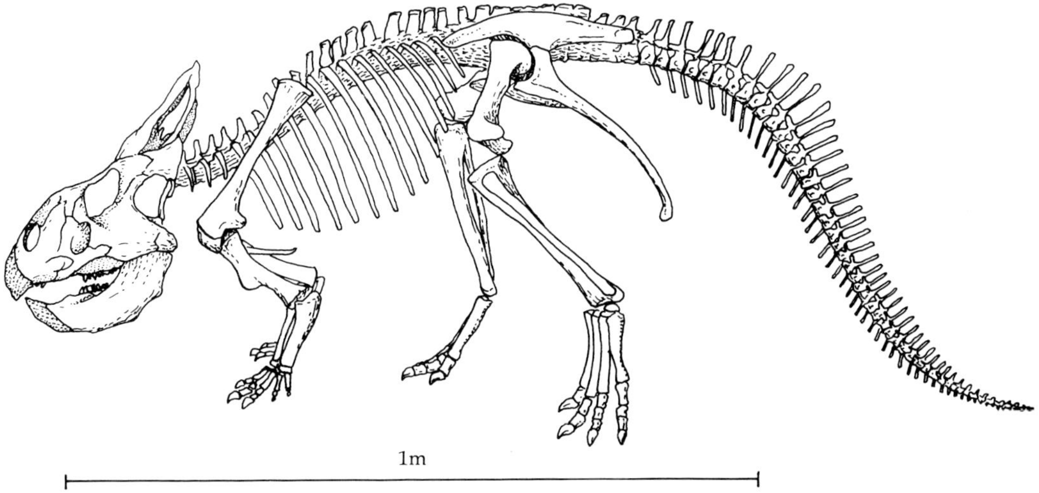

Reconstructed skeleton of Protoceratops andrewsi. *(Based on Sternberg)*

Progressive growth stages of Protoceratops andrewsi *showing a newborn hatchling, a half-grown juvenile, and an old adult. (Based partly on Colbert)*

Protoceratops andrewsi

Protoceratops andrewsi, the best known of the protoceratopians or "proto-horned dinosaurs", is well known from a large number of specimens representing many growth stages from newly emerged hatchlings to fully grown adults. These were collected from the late Cretaceous Djadokhta Formation beds of Bayn Dzak in southern Mongolia, originally by the 1923 Central Asiatic Expedition of the American Museum of Natural History, and later by several Russian and joint Polish-Mongolian Expeditions to the same area.

The adult *Protoceratops* was a small dinosaur, about two metres in length including the tail, but it was a heavily built, stocky animal and had a large parrot-like horny beak and a broad, well-developed bony frill extending at the back of the head. Areas of the skull above the eyes and the snout were raised into irregular protuberances, but the large horns that developed in the corresponding parts of the skulls of large North American ceratopians were not evident. The neck frill of *Protoceratops* was pierced by two large fenestrae or "windows", and served as an attachment for the thick and powerful jaw muscles. These were far bulkier than those possessed by any other herbivorous dinosaur of this size, and indicates an immensely strong bite capability. The tooth arrangement was very similar to those of hadrosaurs, the teeth being small, replaceable and peg-like, and arranged in batteries. However, unlike those of the hadrosaurs, which were used primarily for grinding, they operated more as scissor-like vertical shearing blades, slicing herbaceous food into easily digestible material. *Protoceratops* was ideally suited to feeding upon the fibrous fronds of the many different types of palms that had become common during the late Cretaceous. Their bunched trunks would also have served as a convenient sanctuary from predators for inoffensive plant eaters like the protoceratopians.

There appears to have been little observable sexual dimorphism within *Protoceratops andrewsi*, and both Dodson and Kurzanov have attempted to identify sexual characteristics by biometrical analyses of skulls. Several variables may be relevant in this respect, these being frill width, angle of the frill crest and angle of the frill expansion, nasal height, and facial angles.

Of equal importance to the many *Protoceratops* skeletons was the prolific number of individual and complete nests of eggs found in the same strata. These were the first identifiable dinosaur eggs to be found and analysis of their micro-structure revealed them to be similar to the eggs of birds, though this is probably due to ecological reasons rather than a close evolutionary relationship. The arrangement and number of eggs in the larger nests indicates that *Protoceratops* was a communal nesting dinosaur, because some of the nests contained too many eggs to have been laid by one animal.

Although it was originally believed that the descendants of *Protoceratops* migrated from Central Asia across the Bering Straits to western North America where they gave rise to the "true" horned ceratopians, evidence now suggests that the Bering Straits area was mostly submerged during this time, and a recent study by Polish palaeontologists indicates that rather than being an ancestral type, *Protoceratops* exhibits characteristics that are quite specialised. It now appears that the ancestors of the North American ceratopians were more likely to have been animals similar to the small, fleet-footed Mongolian genus *Microceratops* (= tiny horned-face) that was capable of running bipedally, and that the migration to the American landmass took place before the late Cretaceous. Other Mongolian protoceratopians were *Bagaceratops* (=small horn-face), a small genus from the Khermeen Tsav Formation, and *Breviceratops* (=short horn-face) from the Barun Goyot Formation, an animal similar to *Protoceratops* but with a definite nasal horn. North American genera included *Montanaceratops* (=horn-face from Montana) a *Protoceratops*-like form with a large nasal horn, and *Leptoceratops* (=slender horn-face), a small slender animal from Alberta and Wyoming.

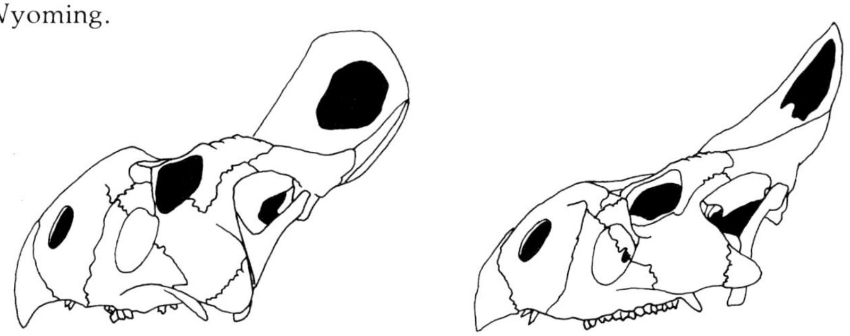

Comparative skulls of a male (left) *and female* Protoceratops. *(Based on Dodson)*

Opposite: Protoceratops andrewsi. *(Based on Z. Burian)*

A well developed skull frill formed by the parietal and squamosal bones covered the neck area and provided an attachment site for the jaw musculature which was capable of powerful compressive forces. The numerous small, peg-like teeth were stacked in tightly-packed tiers, and functioned in a slicing motion to chop and cut herbaceous material. Because of its strong jaws and efficient dentition, *Protoceratops* probably subsisted on a varied diet of low-level leaves, twigs, seeds, fruits, and fibrous palm fronds. Large quantities of such food could be processed because the capacity of the mouth was increased by the presence of pouch-like muscular cheeks. As for the newly hatched *Protoceratops*, they exhibited all the features of the adults, including the sharply-formed faces with parrot beaks and the protective neck frills, but their eyes were comparatively much larger. As stated earlier, the original diagnostic work carried out on *Protoceratops* indicated that this dinosaur lay at the ancestral base of the ceratopian family tree. However, at that time, *Protoceratops* was the only protoceratopian known. Now, other closely-related genera have since been found, and recent work indicates that *Protoceratops* could well be a very specialised form.

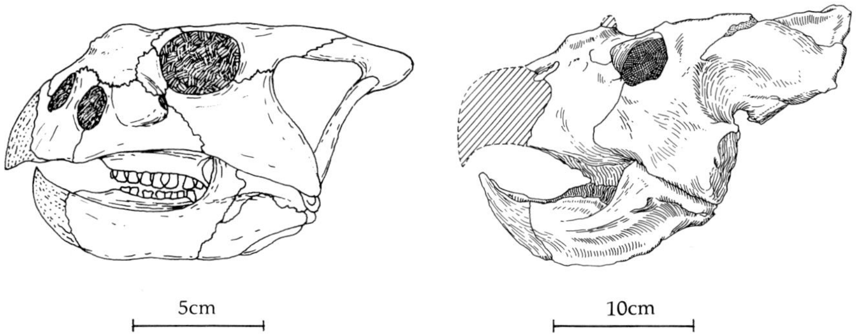

Skulls of the protoceratopians Bagaceratops rozhdestvenskyi *(left) from the Khermeen Tsav Formation, and* Breviceratops kozlowskii *from the Barun Goyot Formation.* B. rozhdestvenskyi *probably had a small nasal horn in life and is known from many skulls representing all age groups (p. 117).* B. kozlowskii, *which was identified more recently (p. 129), was closely related to* Protoceratops andrewsi, *had a small horn, and is represented by a more fragmentary collection of skulls. (*Bagaceratops *based on Maryanska & Osmolska,* Breviceratops *based on Kurzanov)*

Another of the smaller dinosaurs that inhabited this part of Asia somewhat earlier than *Protoceratops* was represented by two reasonably well-preserved skeletons found by the Americans in the Oshih Formation that was early to middle Cretaceous in age. Originally designated as two genera, *Psittacosaurus* and *Protiguanodon*, they are both now recognised as specimens of the one genus, *Psittacosaurus* (= parrot reptile). This was a two-metre-long, quite lightly-built and slow dinosaur that was at first thought to be an ornithopod related to *Hypsilophodon*. Although it usually progressed on its blunt-clawed, strong hind legs like an ornithopod, *Psittacosaurus* was apparently also able to walk quadrupedally. The part of the body that revealed *Psittacosaurus'* important relationship to the ceratopian dinosaurs is the parrot-like skull which was strongly constructed and blunt-snouted, and terminated in a toothless, ceratopian-like beak. At the back of the skull was the faintest hint of the beginnings of a frill.

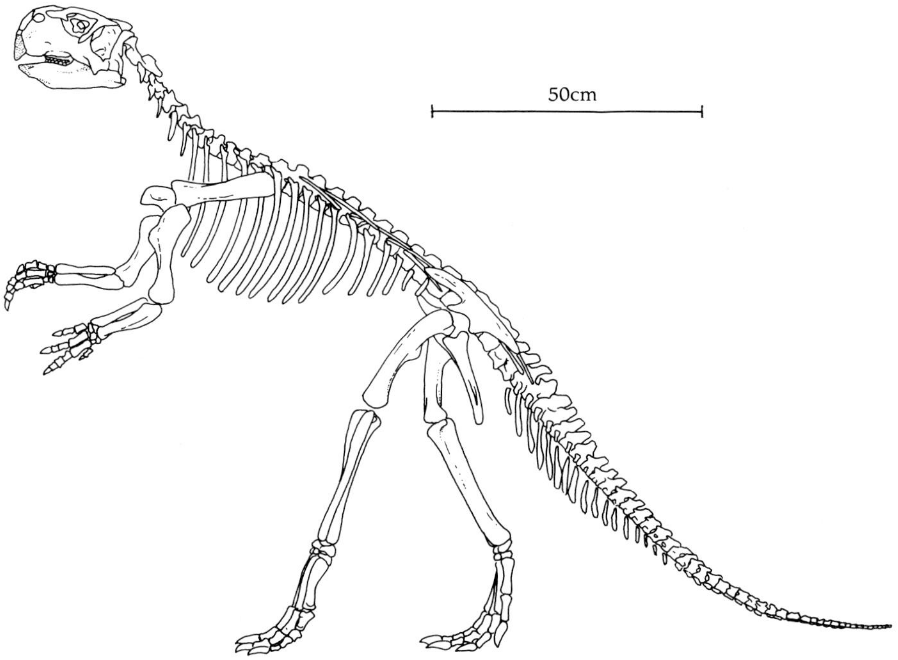

Reconstructed skeleton of Psittacosaurus mongoliensis. *(Based on Osborn)*

Some palaeontologists believe that the appearance of the psittacosaurs coincided with the evolution of the angiosperms, and that the parrot-like beak of these dinosaurs enabled them to cope more easily with changing herbaceous diets. The mixture of ornithopod body morphology combined with ceratopian-like skull features indicates that the small *Psittacosaurus* was the earliest and most primitive of the ceratopians. In contrast, *Protoceratops* and other protoceratopians, namely *Bagaceratops* and *Montanaceratops*, show stronger trends toward the larger "true" ceratopians of North America such as *Styracosaurus*, *Triceratops* and *Torosaurus*.

In 1980, the American palaeontologist Walter Coombs Jr. found some further psittacosaur material among the fragments collected on the 1922-1925 American Museum Expeditions. What is most interesting about the two skulls and partial skeleton that he identified was their small size. The skulls measured just forty-two and twenty-eight millimetres, suggesting that the full length of the juveniles to which they belonged was only forty and twenty-five centimetres respectively. Like the young hatchling *Protoceratops*, these tiny psittacosaurs also exhibited all the morphological features of their parents, and again, showed eye sockets comparatively much larger than in the adults. There were obvious signs of teeth abrasion on these two young hatchlings, and this could indicate that either they were very precocious and already foraging for themselves, or what is also possible, that the parents brought food to the hatchlings when they were still in the nest.

Some of the small carnivorous dinosaurs with which *Protoceratops* shared its ancient world in the late Cretaceous included three genera of small, bipedal theropods, *Saurornithoides*, *Velociraptor* and *Oviraptor*. The scattered remains of *Saurornithoides* (= bird-like reptile) were all collected from a sandstone matrix in 1923 by a Chinese assistant, and consisted of parts of the skull, backbone, pelvis, legs, and feet. The narrow-muzzled and bird-like skull had very large orbits whose position indicated a degree of binocular vision. The brain capacity was very generous and compared well with that of a modern bird. Indeed it was originally believed that the remains were avian in origin. The finely-built upper jaws contained over thirty teeth, these being unusual because only their back edges were serrated. From these and more recently collected remains, we know that *Saurornithoides* had a lightly-built body, with long, slender legs and arms. The second digit of the foot was armed with an enlarged claw that was probably held well clear of the ground when walking or running.

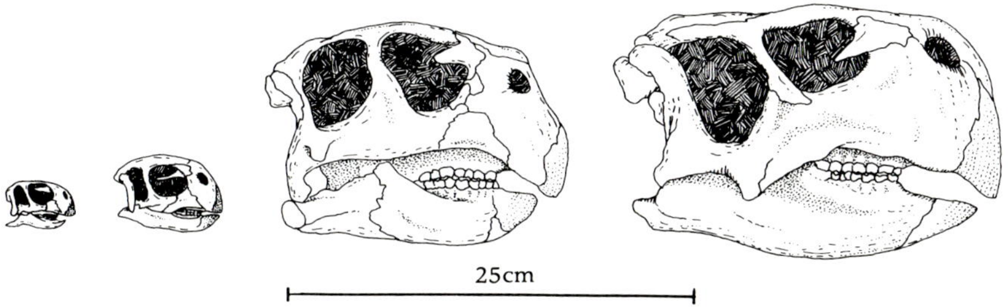

Progressive growth stages of Psittacosaurus *skulls showing a newborn hatchling, a one-year-old juvenile, an adolescent, and an adult.* (*Based on Benton*)

Complete skeleton of Psittacosaurus mongoliensis *from the Lower Cretaceous Barun Bayn Formation of Khamrin Us, as mounted at the Moscow Palaeontological Museum.* (*Courtesy of S.M. Kurzanov, Russian Palaeontological Institute*)

Protoceratops hatchlings

Among the material collected during the American Museum of Natural History Expeditions to Mongolia was a large number of dinosaur eggs found at the celebrated "Flaming Cliffs" locality of Bayn Dzak (Shabarakh Usu in the American literature). There are several types of eggs in the Bayn Dzak strata, attesting to the fact that different species of dinosaurs nested in close proximity to each other. The majority of eggs, however, appear to be of the same type, and these have been assigned to *Protoceratops andrewsi*, itself known from a great number of specimens. The twenty-centimetre-long, oblong-shaped eggs superficially appear to be not unlike those of some modern lizards, and have a rough, corrugated texture. Many of the eggs are grouped together in nest-like arrangements, having been laid in concentric circular patterns. As to why dinosaurs should have favoured the Bayn Dzak locality for nesting is explained by the geology and geography of the area. The quality of the sand must have been suitable for protection of the buried eggs as well as being porous enough to permit heat and gaseous exchange. The area must have been relatively sheltered and although thought to have been somewhat arid, the remains of turtles and small crocodiles show that Bayn Dzak was well watered, probably from shallow lakes to the south.

A number of smaller predatory dinosaurs that were contemporaneous with *Protoceratops* during the late Cretaceous may have been opportunistic nest raiders, plundering the eggs when the adult *Protoceratops* were not in the vicinity. Such dinosaurs included the genera *Velociraptor*, *Oviraptor*, and *Saurornithoides*, though none is considered to have specifically been an egg-eating predator. Today, just as during the Cretaceous, the nests of many reptiles are similarly at the mercy of other reptiles, and the eggs of marine turtles and crocodiles are often dug up and fed upon by rapacious monitor lizards, even at the risk of being caught by a guardian mother crocodile. Adult *Protoceratops* were not large animals, but neither were they totally defenceless. Their powerful bite provided some protection against predatory dinosaurs, and they probably also chose to live among groves of closely growing trees and low, thorny scrub. This enabled them to carry out their nesting activities in relative safety and seclusion, though it is not known whether the nests were guarded as they are by crocodiles, or if they were abandoned after laying.

It is possible to recreate the ancient environment in which *Protoceratops* congregated and reproduced over 75 million years ago. The Bayn Dzak area was crisscrossed by small streams running between gently sloping sandy areas covered in low, jungly scrub and small groves of palms. Living amidst the vegetation of this tranquil habitat were small communal groups of the slow-moving and passive *Protoceratops*. During the egg-laying season, the females congregated at the same nesting site used in previous years, such as a warm, sheltered and secluded area of loose sand. A female would begin by excavating a shallow depression with her clawed feet, in much the same manner as crocodiles still do. She would then proceed to lay the first inner circle of eggs in the centre of the nest, depositing them one by one, broader end outwards, as she slowly turned full circle. Possibly a further circle outside of this first one would be started, depending on the number of eggs carried. When the laying was completed, she may have adjusted the position of the eggs with her beak. Later, another female approached and continued adding to the outermost circle until she too had completed her laying and adjustment of the egg's positions. After the depression was full, which was sometimes only after thirty or more eggs had been deposited, the nest was covered over with sand, probably by kicking backwards with the hind feet, and left to be incubated by the heat of the sun. Within their elongated capsules, the tiny embryos were miniature replicas of their parents, but with relatively much larger eyes and smaller and flatter neck frills. By the time they reached the length of about twenty-five or thirty centimetres, the young *Protoceratops* were strong enough to break out of their shells using their beaks, struggle free of the nest, and crawl into the sunlight to begin lives of their own.

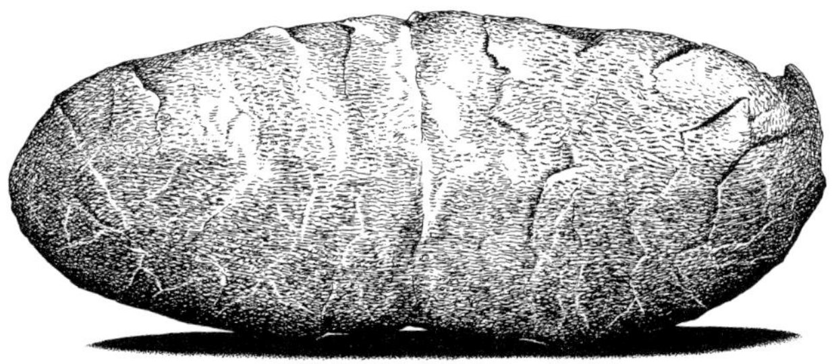

Detail of Protoceratops *egg.*

Opposite: Protoceratops andrewsi hatchlings. *(Based on B. Silverman)*

JKaras '86

This small dinosaur was obviously well-suited for very active predatory behaviour, and a large portion of the brain was dedicated to the fast and smooth coordination of such activity. Because of its large orbits, *Saurornithoides* was likely to have been a nocturnal or twilight predator, exploiting those periods when most other predatory dinosaurs were unable to hunt effectively in the poor light. Of course, these were the times that the small nocturnal Mesozoic mammals would also have ventured out in order to forage. It has been suggested that with its binocular vision and quick reflexes, the fast-moving *Saurornithoides* may have been one of the few dinosaurs to have preyed upon these early mammals, capturing them with its dexterous, grasping fingers. It was into a dangerous world indeed that the earliest of our ancestors ventured during those twilight hours of the Cretaceous nights.

Up until recently, *Saurornithoides* was classified as a dromaeosaur, but in 1974 the Mongolian palaeontologist Rinchen Barsbold described well-preserved, new material of this genus from the Bugeen Tsav area. Because of the presence of certain swollen areas called "parasphenoid

The large-eyed Saurornithoides *appears to have been a twilight predator, perhaps even using its long, nimble fingers to prey upon the shrew-sized late Cretaceous mammals which were nocturnal in habit.* (Based on J. Sibbick)

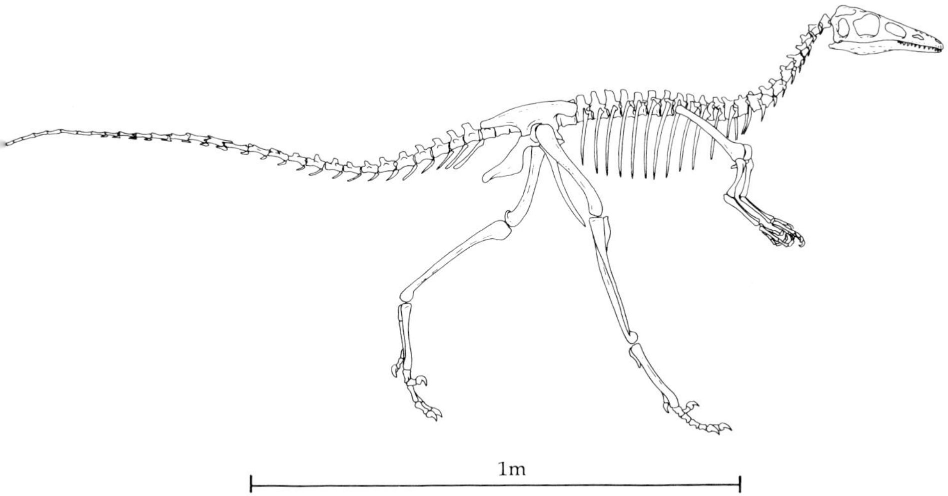

Reconstructed skeleton of Saurornithoides *sp. (Based on K. Carpenter)*

capsules" in the floor of the braincase and in the ear region, as well as several other osteological features, he was able to show that this genus together with the Canadian genus *Stenonychosaurus*, should constitute a family of their own, the Saurornithoididae.

Another of the small theropods was *Velociraptor* (= fast-running robber), a fleet-footed dromaeosaur from the Djadokhta Formation (the same strata that contains *Protoceratops*) and whose remains are also known from Kazakhstan. About 1.8 metres in length, this graceful dinosaur had a long, slender muzzle equipped with sharp, backward-pointing teeth serrated on both front and rear edges. The three-fingered hands had long and mobile fingers, while the legs were slender but powerful. As in *Saurornithoides*, the second toe was modified into a large, sickle-shaped claw that was probably used as a weapon during hunting. *Velociraptor* is considered to have been a very capable predator that may have hunted in small packs, running down dinosaurs much larger than itself. That this predator also came into conflict with smaller dinosaurs like *Protoceratops* was shown by a remarkable find made on the 1971 Polish-Mongolian Expedition (see chapter five). Some unusual bird-like osteological features of velociraptors such as the arm-folding mechanism, the large breastplate, and the peculiar ossified tail rods appear very similar to those found in birds and pterosaurs. This led Gregory Paul to suggest that animals

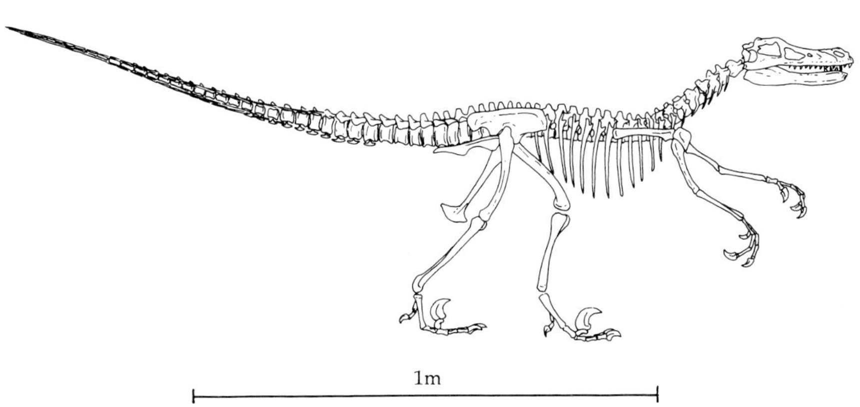

Reconstructed skeleton of Velociraptor *sp.*

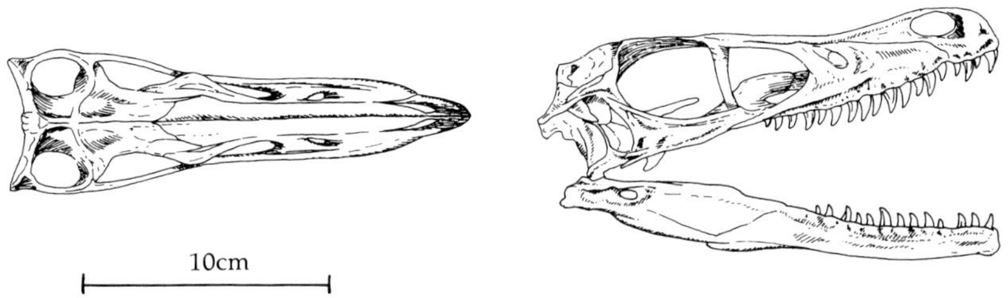

Reconstructed skull of Velociraptor mongoliensis. Left: *Top view.* Right: *Side view. (Based on G. Paul)*

like *Velociraptor* may in fact have been the Cretaceous flightless descendants of Jurassic reptiles such as *Archaeopteryx*. And given the poor flying ability of *Archaeopteryx*, this assumption would appear to be quite plausible.

Also from the same formation as *Velociraptor* is the small, two-metre-long predator *Oviraptor* (= egg-thief), the founding member of the Oviraptorosaurs. *Oviraptor* was initially described from a partial skull and skeleton collected by Olsen in 1923 at Bayn Dzak. This small reptile had a short-snouted skull that was almost toothless as in most ornithomimosaurs, and its jaws terminated in a horn-covered beak that was stronger and deeper than that of an ornithomimosaur. Later described by Osborn, the partial skeleton of *Oviraptor* revealed three elongate and sharply-clawed fingers on the hand, their structure showing similarities to those of the coelurosaur genus *Ornitholestes*, as

Velociraptor and *Oviraptor*

One of the important types of small carnivorous theropods unearthed during the American Museum's Expedition to Bayn Dzak in 1924 was *Velociraptor mongoliensis* (= fast-running robber). There are several species of *Velociraptor* now known from Central Asia and North America, and of these the jackal-sized *Velociraptor mongoliensis* is considered to be the most advanced, even though it was smaller than some other species which reached the size of wolves. Velociraptors were a highly successful group of fleet-footed and lithe predators, remaining virtually unchanged for a period of 50 million years. Along with the North American genus *Dromaeosaurus* and other Mongolian genera known from fragmentary remains, they are classified as dromaeosaurs.

Velociraptors had long, narrow skulls that were up-curved, very graceful, and armed with rows of tightly-packed and bladed teeth of modest size numbering about forty in the upper jaw. The angle of the orbits indicates good binocular vision, and the brain cavity was generous considering the body size. The arms, which could be folded into the torso in a manner similar to the wing-folding mechanism of a bird, had large hands with three long, dexterous fingers. The second toe (the innermost toe in contact with the ground) was hyperextendable and bore a transversely-flattened giant claw that was raised when running, and it is believed that these claws were used as weapons for bringing down prey. Velociraptors may have hunted in gregarious wolf-like packs, utilising their sharp claws to disable larger herbivorous dinosaurs. A clue as to how they might have attacked other animals is provided by the unique fossil of a *Velociraptor* in combat with a *Protoceratops* which shows that the sickle-claws were used as raking weapons to tear at the stomachs of the prey. Even though dromaeosaurs were generally small, with the largest reaching about 3.3 metres in length, on a power to weight ratio they were probably the most lithe and powerful of any known carnivore.

Another small theropod originally found in the Djadokhta Formation of Bayn Dzak during the 1923 American Expedition was *Oviraptor philoceratops* (= egg thief, fond of ceratopian eggs). This was a small dinosaur up to two metres in length, and it had a parrot-like skull with a deep, toothless beak. More remains of oviraptorosaurs have been found by joint Polish-Mongolian Expeditions during 1970 and 1971, and by Russian-Mongolian Expeditions in 1972. These finds show that there was a great degree of variation within oviraptorosaur skulls; while some had well-developed crest-like structures on the top of the snout, others lacked these altogether. As in the case of hadrosaurs, the differences are likely to be due to sex or age.

The Mongolian palaeontologist Rinchen Barsbold and the Polish palaeontologist Halszka Osmolska have shed some light on these poorly-known dinosaurs. Apparently, the body of oviraptorosaurs differed little from those of other small theropods except for the presence of fused collar bones that resembled the corresponding bones in birds. Although the skull had a large number of openings in its sides and appears weak in construction, the jaws were nonetheless capable of powerful biting forces, while their action produced an unusual back-and-forth motion. Situated in the roof of the mouth were two small conical teeth which are not found in any other theropod. *Oviraptor* skulls show great similarity to a bird-like skull found in late Cretaceous strata in Canada. This was described as an unusual ancestral bird named *Caenagnathus* (= recent jawless) in 1940, but it now appears that these and other similar remains probably belong to North American oviraptorosaurs.

The first *Oviraptor* skeleton that was found appeared to be lying over a *Protoceratops* nest as if attempting to steal the eggs. *Oviraptor* probably did not habitually feed on dinosaur eggs, though it may have frequented the Bayn Dzak area to feed on vegetation, pursue small prey, or scavenge on carrion around the dinosaur nesting grounds. If nests were left unguarded for some time, *Oviraptor* might well have tried to uncover the eggs with its dexterous fingers.

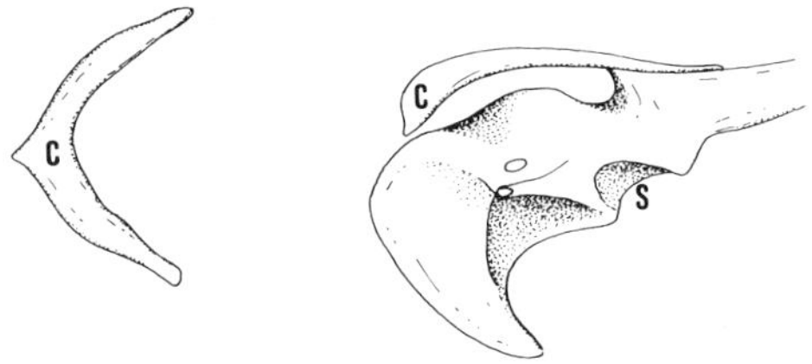

Collar bone C of Oviraptor *in top view* (left) *and side view as attached to pectoral girdle above shoulder socket S. (Based on Barsbold).*

Opposite: Velociraptor mongoliensis (left) *and* Oviraptor philoceratops. *(Based on J. Sibbick)*

well as to dromaeosaurs like *Velociraptor.* Also of importance was the preservation of what appeared to be a bird-like modified collar-bone in the shoulder.

The location of the *Oviraptor* skeleton found by Olsen was interesting in that it lay sprawled over a group of *Protoceratops* eggs. This raises the question of whether or not the *Oviraptor* was caught in the act of attempting to plunder a nest by sucking on the egg contents as is suggested by its specific name *philoceratops.* Osborn assumed that the small animal may have been overtaken by a sandstorm and buried while robbing a nest, while its somewhat crushed skull could have been the result of being bitten by an adult *Protoceratops* which may have surprised the egg-thief. More recently found and better-preserved skulls of *Oviraptor* show that the jaws had a very distinctive movement and were capable of powerful compressive forces far in excess of what would be required to crush eggs. So while *Oviraptor* may have been an opportunistic scavenger of dinosaur eggs, it is doubtful that these constituted a main part of its diet. Regular food was more likely to have been low-level vegetation supplemented by small prey, carrion, and possibly even molluscs.

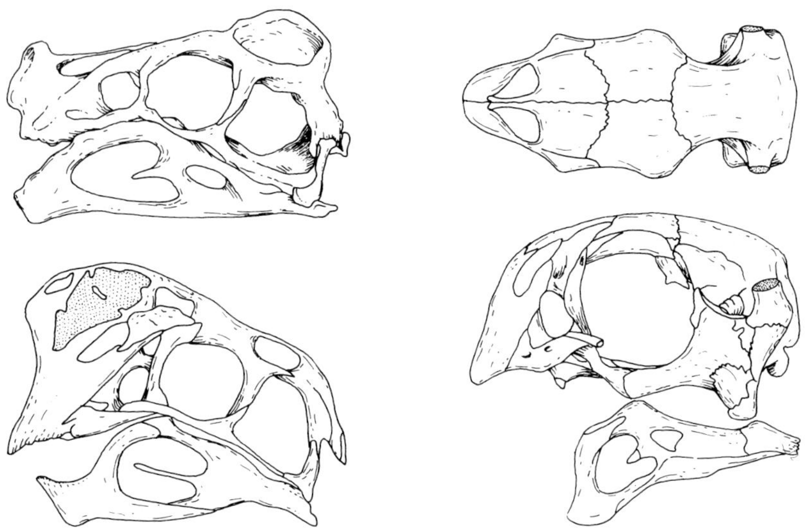

A great amount of variation can be seen between the skulls of Mongolian oviraptorosaurs. The two skulls at left shown here have both been assigned to Oviraptor philoceratops, *though the top skull which was found during the American Museum Expeditions, lacks the large nasal crest so evident in the skull beneath which was found more recently by Mongolian scientists. The two skulls at right are top and side views of a new species found on the joint Polish-Mongolian Expeditions. The relatively large orbits and smooth contours indicate that it is a juvenile. All skulls are shown with snout to the left and are not drawn to scale. (Top left based on Osborn, lower left based on Barsbold, and right two based on Osmolska)*

After the completion of the American Museum's 1923 season's field work, Andrews returned to the United States to find that news of the dinosaur eggs had preceded him and had already reached an incredulous public. That dinosaurs had laid eggs no doubt seemed acceptable and was even expected, but that such fragile objects could have been preserved for what was then believed to have been a span of 10 million years seemed fantastic (this was well before the time of accurate radioactive dating!). Docking at Seattle, Andrews's ship was immediately swamped by newspaper reporters offering grand sums for exclusive rights to photographs of the prehistoric objects, and very soon, dinosaur eggs became world news transmitted to every corner of the globe. One of the eggs was put up for auction as a publicity stunt, fetching a price of $5000 from Austin Colgate, a well-to-do uncle of the expedition's mechanic, Bayard Colgate. Publicity soon earned Andrews over $270,000 which would go towards covering the costs of the next (1925) expedition. But the publicity was not all favourable. News of the assumed value of the eggs also spread to China and Mongolia at a time of deep xenophobia. Officials there, already suspicious of the intentions of the American Museum, and believing that valuable "stone eggs" were being systematically removed from the Gobi via China, demanded that government representatives accompany any future American expeditions.

3.4. Mammals from the Dunes

The 1925 American Museum Expedition would be the largest in the series, and as expected, much of the field work was centred once more at the "Flaming Cliffs" locality. This time, the expedition members were searching not only for dinosaurs, but for Cretaceous mammal remains. One of the fossils that had been recovered from Bayn Dzak in 1923 and sent on to New York was found to be the tiny skull of a shrew-like mammal that lived during the time of *Protoceratops.* In 1925,

Mesozoic mammals were just as much a palaeontological scarcity as dinosaur eggs, and from the American Museum in New York, William Diller Matthew wrote urging Andrews to search for more skulls. Eventually, ten incomplete skulls and isolated jaws of Mesozoic mammals were recovered from Bayn Dzak, the largest skull no more than fifty-five millimetres long. The entire collection could fit into a cigar box, yet these finds were just as, if not more important than the dinosaur remains.

These were all that remained of the various Mesozoic mammals that had been contemporaneous with the dinosaurs, patiently awaiting their late Cretaceous demise before being able to rapidly diversify and refill some of the ecological niches vacated by the reptiles. Up until the 1920s, only several teeth and jaws of Mesozoic mammals had been found (at that time, the skull of the genus *Trytilodon* from the South African Triassic was thought to be a multituberculate mammal and as such was a much treasured possession of the British Museum, but this genus has subsequently been referred to the reptilian suborder Cynodontia). The American Museum finds included two multituberculate skulls (the multituberculates are now extinct), but more significantly, they also included eight skulls belonging to insectivores, the first skulls of placental mammals ever to be found in Cretaceous deposits. Later, in 1926, Gregory and Simpson placed them in four new genera, *Zalambdalestes*, *Deltatheridium*, *Hyotheridium* and *Deltatheroides*.

Such finds confirmed that by the late Cretaceous, the placental mammals were already quite separate from the marsupials (it is now believed that the placentals did not evolve from the marsupials, but simply shared the same original ancestor), while Osborn's hypothesis concerning the Central Asian origin of the placentals appeared to have been borne out. But in subsequent years, many other placental mammals were uncovered in Inner Mongolian, North American, and even European (French) deposits, some of which were from strata as old as Lower Cretaceous. In 1989 however, Zofia Kielan-Jaworowska (leader of the joint Polish-Mongolian Gobi Expeditions) and Demberlyin Dashzeveg described some thirty-nine new placental specimens from the Khovboor locality of Mongolia. Within this material were the remains of the genus *Prokennalestes* which at the time of writing appears to be the oldest placental mammal yet found.

The American Museum Expeditions also recovered much larger mammals from post-Mesozoic strata, at localities both within Mongolia as well as just outside its borders within Inner Mongolian territory which was by then being administered by China. From Central Mongolia came fragmentary remains of a giant hornless rhinoceros previously known from Kazakhstan and Pakistan (see next chapter). Among the more unexpected finds was *Andrewsarchus mongoliensis*, unearthed during 1923 in late Eocene beds (about 43-38 million years ago) at the Irdin Manha locality. *Andrewsarchus* represented an intermediate stage between the flesh-eating mammals or creodonts (= flesh-toothed mammals) and the vegetarian condylarth mammals (= knuckle-jointed mammals), both of which are thought to have originally evolved from the small late Cretaceous insectivores. *Andrewsarchus* was a large animal, about four metres in length but its most impressive feature was the huge eighty-four-centimetre-long skull, far larger than that of any mammalian carnivore, either extinct or extant. For comparison, this is four times longer than the skull of a wolf, or double that of the Alaskan brown bear (the largest modern

The giant carnivore Andrewsarchus mongoliensis *from the late Eocene. (Based on Z. Burian)*

carnivorous land-mammal). The flat-sided muzzle carried a fearsome complement of teeth, including giant, carnivore-like incisors, but the joints of the four-toed feet were similar in some aspects to the hoofs of primitive ungulates. Like modern bears, *Andrewsarchus* probably had an omnivorous diet, but preferred meat when available.

Comparable sizes of the skull of Andrewsarchus mongoliensis *with that of a wolf* Canis lupis. *Based on Andrews)*

After the third Central Asiatic Expedition of 1925, changes taking place within Mongolia as well as the unstable political climate in neighbouring Peking dictated that future expeditions would be restricted to working outside of the areas where the main dinosaur fossils had been found. Apart from Bayn Dzak and Iren Dabasu, these included the localities of On Gong, Oshih, and Ondai Sair, areas that like Bayn Dzak and Iren Dabasu, were eroded, natural basins separated from each other by desolate tracts of desert. There was another expedition organised in 1926 to Yunnan Province, and a further one in 1928 during which Inner Mongolian sites were explored and worked. But by this time, xenophobia in China was intensifying and at the conclusion of the 1928 season's field work, that year's collection of some eighty-seven crates of fossils was temporarily confiscated. It was eventually released only after protracted negotiations between the Chinese government and the Committee for the Preservation of Ancient Objects on one side, and the American State Department and the American Museum on the other.

After a year's delay due to objections from China, Andrews was permitted to return to Inner Mongolia in May of 1930, and this expedition would prove to be the last in the series. It was decided to concentrate on an area just south of Mongolia's border with Inner Mongolia, at a place called "Wolf Camp", so named because of the large, voracious wolves that preyed on the sheep and goats of a local Mongol village. In two months, thirteen were shot by the expedition members. Here, at a brief stay in the autumn of 1928, a jaw of the shovel-toothed mastodon *Platybelodon* had been found in Miocene deposits (from 11 to 6 million years old). Six weeks of further excavations during the 1930 season revealed large numbers of fragmentary remains of platybelodons of all sizes, from newborn young to adult bulls with skulls over 1.6 metres long. One find even consisted of the remains of an unborn young within the pelvis of a female. Evidently, the "Wolf Camp" area had been a large basin flooded by a shallow lake upon which grew abundant aquatic vegetation. The platybelodons came here in large numbers to feed by dredging up masses of lake weed in their huge shovel-shaped lower jaws which terminated in two large, chisel-shaped incisors. The upper jaw was also unusual, there being no trunk proper, but rather a highly muscular, thick lip that was probably used to help rake the food into the lower jaw.

In terms of bulk, 1930 was the most productive year for Andrews and his co-workers. Some ninety-one cases of fossils (mostly of Tertiary mammals) were the result of that season's work, a fitting conclusion to the American Museum's Central Asiatic Expeditions. Up to that time, these had been the largest and most expensive series of land-based explorations ever organised by the United States. Andrew's

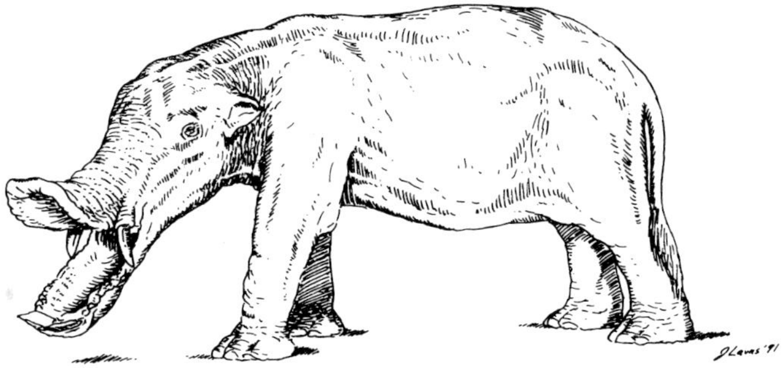

The shovel-toothed mastodon Platybelodon grangeri *from the late Miocene. (Based on an American Museum model)*

original estimate of $250,000 had spiralled to well over $700,000, yet everyone concerned realised that the expense had been more than justified. Not only had a new chapter in vertebrate palaeontology been opened, but a large part of Mongolia had been mapped for the first time, and thousands of botanical and zoological specimens had been collected. In addition, excavations had been conducted at many archaeological sites that while not providing the hoped-for remains of ancestral hominids, had nonetheless yielded important artefacts from Palaeolithic and Neolithic cultures.

These results had been achieved in the face of great difficulties, both environmental and man-made. In the 1920s, that part of Asia was a land of civil wars, marauding gangs of bandits, and soldiers who were a law unto themselves. Caravans were attacked with regularity, their occupants frequently executed. During the winter of 1922, expedition member Charles Coleman had been fatally shot by Chinese soldiers at the border near Kalgan on a particularly notorious part of the road. And later, while on a return trip to Kalgan, Andrews had to break up an armed ambush by driving his field car at full throttle directly at three mounted robbers. Finally, just two days after the last expedition

In search of water, assistant Merin leads a camel caravan across the desolate dunes of Tsagan Nor. (Photo by J.B. Shackelford, courtesy of department of library services, American Museum of Natural History)

safely returned to Kalgan with its load of fossils, bandits took control of the area they had just passed through, causing a halt to all further caravan traffic.

Yet in spite of such conditions, the expeditions were not without their anecdotal, light-hearted moments. Andrews recounts how upon returning to Kalgan from Urga one season, it was discovered that the oil from the car motors had leaked out along the way due to the rough terrain. Mongols from a nearby caravan had a supply of mutton fat, so this was boiled and poured into the engines. Andrews and his colleagues, who had not eaten for some time, proceeded on their way but for the rest of the journey had to endure the tantalising odour of roast lamb emanating from the warm motors.

Even after reviewing the impressive results of the Central Asiatic Expeditions in 1930, Osborn proposed that they had still barely scratched the surface of the Gobi sands, and that there were many

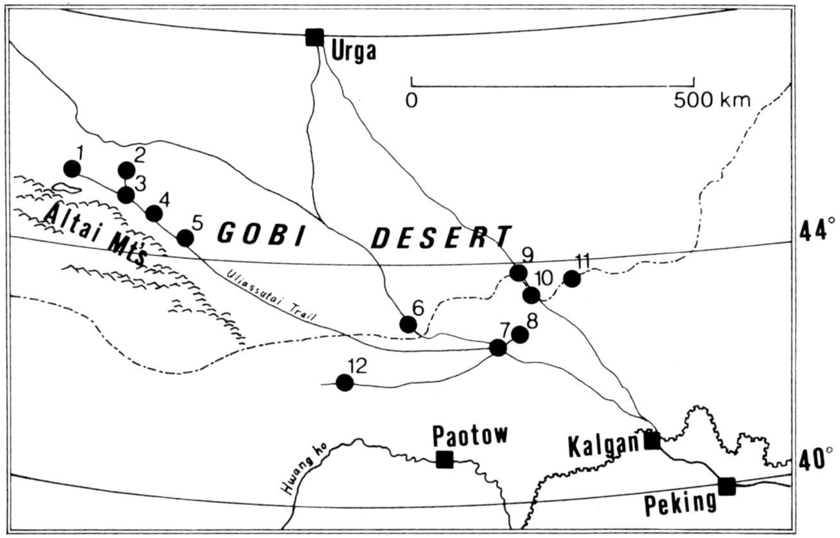

Map of areas explored by the American Museum of Natural History's Central Asiatic Expeditions between 1922 and 1930. Remains of Cretaceous dinosaurs were excavated at 4) Oshih Basin, 5) Bayn Dzak, 9) Iren Dabasu, and 12) On Gong. (Based on Osborn, 1930)

important gaps yet to be filled in the Mesozoic history of the area. Only large-scale future expeditions would be able to find and work major new dinosaur-bearing strata of early and late Cretaceous age, strata that he was sure contained much larger dinosaurs than had been found by the Central Asiatic Expeditions.

Andrews was somewhat less optimistic than Osborn, having written as early as 1926 that the American Museum Expeditions had probably exploited the most important fossil localities of Mongolia. When he wrote this, interest in palaeontology, and particularly palaeoanthropology, had begun to spread to other areas of the world, one of the main reasons being the finding of the *Australopithecus africanus* skull in southern Africa by Raymond Dart in 1924. This was the first hint that the American Museum Expeditions may have been looking for the origins of Man on the wrong continent. The preliminary report by Dart issued on February 7 of the following year had stated simply but very prophetically "....the specimen is of importance because it exhibits an extinct race of apes intermediate between living anthropoids and man...". Later finds continued to focus the attention of anthropologists more and more toward the Dark Continent as the place in which to search for Man's immediate ancestors. In any case, the sensation surrounding the discovery of the Bayn Dzak dinosaur nests had already totally eclipsed any unfulfilled palaeoanthropological aims of the Central Asiatic Expeditions.

After the 1930s, and particularly by the time World War II had erupted, the practice of mounting labour-intensive forays to remote areas of the world such as Andrews had so brilliantly done, was no longer considered feasible in the United States. Andrews was later to serve seven years as the director of the American Museum of Natural History, but having spent most of his working life in the field, the desk-bound job left him increasingly frustrated and restless. Eventually retiring from the museum staff in 1941, he died nineteen years later at the age of seventy-six. A keen and vivid writer as well as an enthusiastic zoologist and explorer, Roy Chapman Andrews has left a rich legacy of popular literary works. Among them is his 1926 narrative *On the trail of ancient Man*, a fascinating account of the pioneering adventures of the Central Asiatic Expeditions in the deserts of Mongolia.

Chapter Four

Road of the Wind

4.1. The Muscovite Campaign

SOME one and a half decades after the 1930 American Museum of Natural History Expedition to the Gobi, palaeontological exploration of the desert was resumed, this time by scientists from a country directly neighbouring Mongolia, the Soviet Union. The political and cultural changes that had occurred in Mongolia since the 1920s placed restrictions on which foreigners were permitted to conduct field work within the country, and the reasons for these changes can be traced back to the Chinese revolution against Manchu rule that erupted in 1911.

At this time, the Mongolians, with backing from Tsarist Russia, also began a struggle for independence of their own, having been under Chinese domination since the 17th century. Consequently, in 1915, Russia and China, the two giants which totally surrounded Mongolia, signed a treaty effectively dividing the country into two parts; Outer Mongolia (now formally known as the Mongolian People's Republic or simply Mongolia) and Inner Mongolia (now the Inner Mongolian Autonomous Region of China). Paradoxically, both Russia and China had once found themselves engulfed in the Great Mongol Empire founded by Ghenghis Khan in 1206. (Just west of the current capital, Ulan Bator, a large stone tortoise marks the site of the old Khan capital, Karakorum, which was razed in 1388 by the Chinese).

In Mongolia, Buddhist rule lasted until the last Living Buddha (Head of State) died in 1924. Up until then, as many as four out of every ten of the country's adult male population were celibate lamas, many of whom owned extensive estates worked by serfs. On July 11th, 1921, just prior to the first of the American Expeditions, a communist regime came to power as a result of a revolution that had begun partly against Buddhist landowners and partly against Chinese rule. Under the new government's directives, Buddhism and the old, traditional nomadic way of life were to give way to socialism and industrialism. The country came increasingly within the Soviet camp of political and cultural influence, and in 1941, the Scientific Commission of Mongolia approached the Soviet Academy of Sciences with a proposition of extensive palaeontological exploration of the portion of the Gobi within its borders. However, the German invasion of the Soviet Union in June of that year and the subsequent outbreak of war intervened, and any large-scale expeditionary plans had to be postponed indefinitely.

By May 8, 1945, the war with Germany was over, but the Japanese campaign in Asia on the eastern side of the Soviet Union was not, and this campaign involved the Gobi area. It was not surprising that the hostilities of World War II had reached to this part of Asia, because in 1931, just one year after the final American Museum Expedition to Inner Mongolia, Japanese forces had occupied that country and installed the Chinese-born "emperor" P'u Yi as a puppet leader of the country that was to be known as "Manchukuo" (Manchuria). Following the Soviet declaration of war on Japan on August 8, 1945, the (Soviet) Red Army and navy forces crossed the borders of Manchuria in two waves, one from neighbouring Mongolia, the other from the maritime provinces of Siberia. In fourteen short days, they had overrun the country and on August 14, Japan surrendered on all fronts.

Following the end of the War, negotiations between Mongolia and the Soviet Union concerning palaeontological ventures resumed in 1946, and later that year, the Russian Palaeontological Institute dispatched a reconnaissance expedition to the Gobi. The object was primarily exploratory: to locate potential fossiliferous localities as well as begin preliminary excavations in preparation for further work in later seasons.

Chief scientific advisor for the Russian Expeditions was the Academician I. Orlov (an Academician being the highest elected position for a Soviet scientist) and then Director of the Soviet

Palaeontological Institute. Orlov was an internationally recognised authority on mammals and reptiles and his involvement with these ventures stressed the degree of importance that the Palaeontological Institute had attached to them. All field work was to be conducted under the guidance and directives of the Academy of Sciences, while the individual expeditions were to be led by a leading Russian expert on extinct animals, Ivan A. Efremov. A notable vertebrate palaeontologist specialising in the reptiles, Efremov was also widely known in the Soviet Union as the author of many science fiction books. Described by his peers as a person with an inquiring and restless mind, he possessed the ideal qualifications for leadership of ventures on such a large scale. The adventures of the dinosaur hunters and the atmosphere of the timeless desert are recorded in his 1956 book, *Road of the Wind*.

Others present on this reconnaissance expedition included six more scientists and technicians including K.K. Flerov, whose convincing recreations of extinct animals are well known. This small group remained in the field for two summer months, within which time they managed to make two traverses, one in the southeastern Gobi where, in the Nemegt Valley, rich accumulations of the remains of medium and large-sized dinosaurs were located. On the basis of the experience gained during this summer work, plans were then laid for a longer and more extensive expedition to be undertaken in 1948.

Preparations for the 1948 Expedition began in the middle of the harsh Russian winter when the expedition members departed Moscow to begin the trans-continental journey to Ulan Bator, arriving at the city from the north. When the Americans had reached Ulan Bator (then called Urga) on their first expedition (arriving from the south), they had found a rather backward and desolate, hardly more than village-sized outpost upon the windswept Asian steppes. In the years since, the industrial and cultural revolution had begun to transform the city into a modern, urban capital of the 20th century. Modern buildings lined the many avenues and parks and the country's first university, the Mongolian State, had recently been founded in 1941.

Naturally, the Mongolians themselves showed much interest in the forthcoming ventures. To them, fossils were not new; but what was new was the explanation of what they were and how they came to be there. Nomadic herdsmen sometimes found curious bones exposed in the desert, and for thousands of years, Chinese apothecaries had utilised fossils or "dragon bones" in the manufacture of potions used to treat conditions ranging from impotence to anxiety. Some cave formations in southern China are so richly endowed with fossils that even today, they are actively mined by local peasants who sell their finds to town apothecary owners.

Second in command to Efremov on this expedition was one of the top Russian authorities on reptile taxonomy, Anatole K. Rozhdestvensky. His lucid account of the Russian Expeditions, *In the footsteps of dinosaurs in the Gobi Desert*, was published in 1957. Also present to note were the palaeontologist E.A. Maleev, and N. Novojilov, the chief geologist. Altogether, fifteen scientific members participated, with a further eight labourers and technicians and six drivers completing the team. The expeditions were provided with a number of heavy-duty ZIL trucks (from the Zavod Imieni Likhacheva truck and limousine factory, Moscow) for the hauling of equipment and supplies, as well as several GAZ field-cars (open-sided, all-terrain vehicles). These cars were suitable for the very rough terrain found in many sections of the Gobi and were ideal for advance scouting and observation work. After some two months of preparation in the Mongolian capital, the small convoy left Ulan Bator on March 18, winding its way slowly south out across the open panoramic steppes.

Participants on the 1948 Russian Academy of Sciences' Expedition to the Gobi (left to right) I. Sidorov, P. Petrunin, N. Brilijv, I. Alexandrov, Namnandordg, J. Eaglon, I. Efremov, I. Orlov (standing), I. Likhatchev, N. Shkilev, N. Vilezhanin, V. Pronin, Damdin. Others not in the photograph are T. Bezborodov, M. Lookijnova, E.A. Maleev, N. Novojilov, V. Presniakov and A.K. Rozhdestvensky. (From the archives of A.K. Rozhdestvensky, courtesy of S.M. Kurzanov, Russian Palaeontological Institute)

4.2. Armoured Dragons

Using the large trucks and smaller, fast field-cars, Efremov and his group were able to penetrate deeper into the desert than any of the American Expeditions. Their work would take them further into the southwestern and western sectors, reaching to the very limits of Mongolia, to areas still totally unaffected by the 20th century. These expeditions naturally had to contend with similar problems to those that had plagued all previous explorers, but made more acute because of the greater distances being covered. The constant shortage of fresh water, the lack of useful landmarks for navigation, the incredibly powerful winds and sandstorms that arose with little warning – blackening out the sun and scattering tents and equipment, and finally venomous vipers, spiders and scorpions, just some of the conditions that had earned the Gobi its unenviable reputation.

Human habitation in many areas of the desert was barely evident; now and then a lonely group of circular yurt tents dotted the monotonous landscape, or nomadic Mongols leading camels or tending sheep might be encountered, but one could travel extensively through the harsher sectors of the Gobi without seeing another human. When the palaeontologists reached the farthest-flung outposts of the country, they found that the arrival of their vehicles often had the unfortunate effect of terrifying the local inhabitants, many of whom had never before seen an automobile. But in the open, treeless steppes, the large trucks proved themselves indispensable in more ways than one: with daytime temperatures often soaring to 40°C, Efremov and his companions often found themselves seeking relief by grabbing their maps and geological charts and planning their next move in the comparative coolness under one of the trucks.

The first excavations were carried out in an area south of Sayn Shand in the southeastern Gobi, known locally as Bayn Shireh. Here, many dinosaur bones lay embedded in the Cretaceous rocks, and skeletons of an interesting new armoured ankylosaur were retrieved from Upper Cretaceous strata. *Talarurus plicatospineus* (*Talarurus* = basket-like, named because the rib cage resembles a basket when viewed from the front) was a large, deep-chested animal that grew from 4.5 to six metres in length, and exhibited the heavy dorsal body armour characteristic of the Ankylosauria. A series of heavy keeled plates almost five centimetres thick provided a formidable yet flexible armour for the back, hips and tail, while along the top and sides of

The desolate landscape of Bayn Shireh, with members of the 1948 Russian Expedition excavating Talarurus *remains from Upper Cretaceous strata at the left of the photograph.* (Courtesy of S.M. Kurzanov, Russian Palaeontological Institute)

Mounted skeleton (minus tail) of Talarurus plicatospineus *at the Moscow Palaeontological Museum. This hippo-shaped ankylosaur was heavily armoured and grew to over six metres in length.* (Courtesy of S.M. Kurzanov, Russian Palaeontological Institute)

the body and tail were arranged rows of many long, sharp spines, providing a valuable defence system against carnivorous dinosaurs. The remains of at least six individual animals and two skulls are known for this species, while more fragmentary material of another species, *Talarurus disparoserratus* was also collected. *Talarurus* is the oldest well known late Cretaceous ankylosaurid, and recent study shows that the genus shows some primitive features subsequently lost in later ankylosaurids, being more characteristic of the nodosaurids, the group from which the ankylosaurids may have evolved.

The group then turned westwards to the Bayn Dzak locality, where the American Expeditions had made such sensational finds in the 1920s. Since that time the red sandstone cliff outcrops had continued to erode, exposing to light many more examples of the small *Protoceratops* skeletons and associated eggs. The Cretaceous strata of Bayn Dzak yielded other dinosaur remains as well, including another large ankylosaurid which was originally named by Maleev as *Syrmosaurus vimicaudus*, but is now recognised as a junior synonym of *Pinacosaurus grangeri*, hitherto only known from a deformed skull collected in the same area by the Americans. *Pinacosaurus* (= board reptile) was a large, primitive ankylosaurid (though more advanced than *Talarurus*) that lived during the late Cretaceous. From 3.5 to five metres in length, it had a relatively slender body that was held well clear of the ground on slender but solid limbs. The long tail terminated in a solid, bony club, and the neck, back and tail were armed with symmetrical rows of pointed plates that were quite separate from each other.

The five-metre-long Pinacosaurus grangeri, *a relatively slender ankylosaur from the late Cretaceous. (Based on G. Olshevsky)*

In addition to the *Talarurus* and *Pinacosaurus* material collected, another large ankylosaur was later recovered from Upper Cretaceous deposits in the Nemegt Valley. Originally designated as *Dyoplosaurus giganteus*, this heavy ankylosaur is now referrable to the genus *Tarchia* established by Maryanska (Tarchia being the Mongolian word for brain, so named because of the relatively high-domed brain-case). *Tarchia* was one of the largest (up to 5.5 metres) and one of the most recent of the Asian ankylosaurs. The broad, triangular head was completely shielded by a patterned arrangement of bony plates like cobblestones, with large spines over and behind the eyes. Smaller, backward-pointing spines adorned the back, as well as the top of the clubbed tail, while a row of long spines ran along the sides of the body, projecting out over the limbs. More material and a similar but new species of this ankylosaur was collected more recently on the Polish-Mongolian Expeditions.

Probable life appearance of the ankylosaur Talarurus plicatospineus *from the late Cretaceous.*

Study of these species has thrown some light on the evolutionary history of this unusual dinosaur group, and its phyletic relationship to the equally unusual stegosaurs (plated dinosaurs), which some experts believe had the same early Jurassic ancestor as the ankylosaurids and nodosaurids. Although ankylosaurids have been found in other localities in Asia and North America, they are comparatively rare fossils. The diversity and quantity of the Gobian genera found during and since the Russian Expeditions is exceptional and indicates that this area was probably a centre of development for the group.

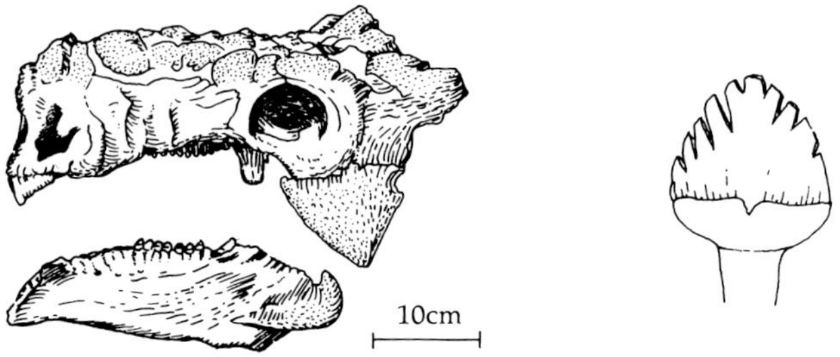

Skull of Tarchia gigantea, *and tooth from the lower jaw (greatly enlarged). (Based on Tumanova)*

Ankylosaurs were heavy and sturdily-built reptiles; the zoological equivalent of a modern battle tank. They appear to have been rather passive and slow-moving in habit, probably frequenting low-lying or rolling steppe areas not too distant from water. In such environments, they could subsist on low- to ground-level plants and they may also have been partly insectivorous. Gobian ankylosaur skulls show a well developed olfactory apparatus indicating that the sense of smell was relied upon to a large extent to locate food and mates, as well as avoid predators. While it has been assumed that they were primarily herbivorous, their hyoid structure is very similar to that of modern snakes and lizards, suggesting the presence of a long and mobile tongue ideally suited to catching insects or other invertebrates and their larvae. The Hungarian palaeontologist Franz Nopcsa surmised that the Canadian genus *Scolosaurus* could have lived on a diet of grasshoppers and beetles, requiring something like 1000 per day. In addition to herbaceous and invertebrate food, ankylosaur diets may have also been supplemented by scavenging on carrion.

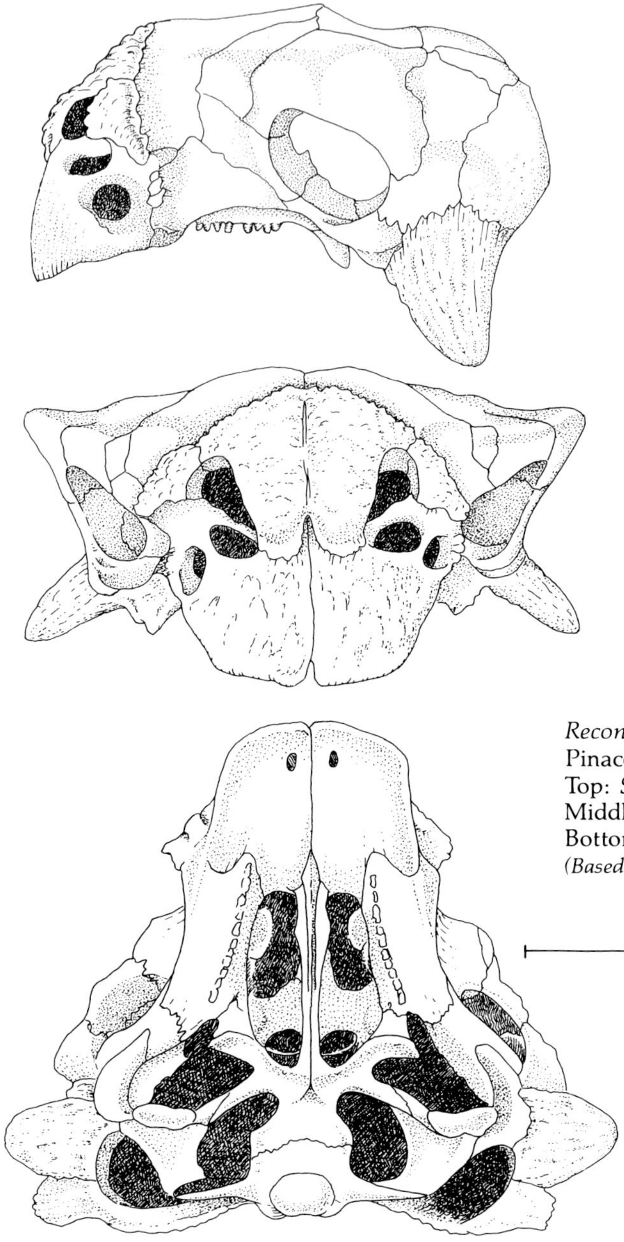

Reconstructed skull of Pinacosaurus grangeri.
Top: *Side view.*
Middle: *Front view.*
Bottom: *Palatal view.*
(Based on Maryanska)

10cm

The classical explanation for the formidable dorsal armoury that these animals possessed has been one of defence against larger carnivorous dinosaurs, who no doubt would have set upon the docile ankylosaurs intent on exposing a weak spot or overturning the beleaguered animals to expose the relatively unprotected belly (though the Mongolian genera do show some ventral armoury). Piercing the heavy armour from the top must have been a near impossible task, even for the most powerful of predators, because every part of the upper body was well protected, including the head. Although bone sutures are still evident in the skulls of juvenile animals, by the time they reached maturity, the bones had completely fused together. Skulls of the genus *Euoplocephalus* show that even the eyes were armoured! Concave, bony eyelid shields could be lowered to protect the eyes against the teeth or claws of an attacker.

The ankylosaurid tail was unique in the reptile world, the two-lobed solid club at the end being formed from bones embedded in the skin becoming fused to the tail vertebrae and to each other. Traces of muscle scars show that the tail was able to be swung from side to side by powerful groups of muscles linking the sides of the tail with the pelvis and hindlimbs. One set, the M. caudi-femoralis longus, ran the entire length of the tail, attaching to the end at the heavily ossified part of the club. All in all, the tail was a powerful defensive weapon; all the more effective when one considers that the predatory dinosaurs were bipedal, and therefore more liable to be toppled than quadrupeds.

Another tactic of defence would have involved crouching low to the ground, tortoise-fashion, with the feet tucked in under the protection of the armour. The great weight (up to two tonnes in some species), and the very low centre of gravity would provide good stability while the tail could still be used to make an occasional counter-swipe. Then it became a life-and-death game of perseverance while the attacker pitted tooth and claw against its stubborn adversary.

Ankylosaurs may have had other behavioural adaptations to avoid predators. Maleev suggested that *Pinacosaurus* might have been able to dig out shallow depressions in the ground, using the spines to rake away the sandy soil. When lying in such "burrows", the only parts of the body exposed would be the impregnable armour. The shape of some ankylosaurs, especially those with flattened bodies protected by keeled plates along the sides, and the presence of the bony eyelids, certainly appears to support this view.

Over 60 million years after the last of the ankylosaurs had become extinct, a group of Pleistocene mammals, the glyptodonts, which were the ancestors of modern armadillos, again evolved ankylosaur-like

The formidable armour of Talarurus *might have proven impregnable, even against the largest carnivorous dinosaurs.*

Ankylosaurids

The Gobi Desert accounts for most of the well preserved ankylosaurid remains found to date. These include (in order of chronological first appearance in the fossil record, *Shamosaurus* and *Sauroplites*, *Talarurus* and *Amptosaurus*, *Pinacosaurus* (= *Syrmosaurus*), *Saichania*, and *Tarchia* (= *Dyoplosaurus*). In addition, several probable new Gobian genera are presently being described. Remains of these genera have been found on the Russian Academy of Sciences' Expeditions (1946–1949), the joint Polish-Mongolian Expeditions between 1963 and 1971, and during more recent joint Russian-Mongolian field work, while the taxonomy has recently been clarified by Teresa Maryanska of the Polish Institute of Palaeobiology.

The most heavily armoured of all the dinosaurs, ankylosaurs were largely non-gregarious, deliberate moving and passive reptiles that relied primarily on their flexible dorsal body armour for defence against the many contemporary carnivorous dinosaurs. (This same strategy of protection against predators was evolved again much later in a group of South American mammals, the Pleistocene glyptodonts, who developed a similar though non-flexible tortoise-like bony carapace over the back, as well as clubbed tails). The skulls of ankylosaurs are so heavily armoured that the individual bone sutures of adult animals are difficult to recognise, while even the small eyes had the protection of bony eyelid covers. The three genera shown opposite are, from left to right, *Tarchia*, *Saichania*, and *Talarurus*, and though they do not all appear in the same fossil horizons, are depicted together here to show the variety of armour evolved in these animals.

Tarchia, at about 5.5 metres length, had a particularly heavily armoured, spike-adorned skull, with many short spines set in rows over the back, and larger spines projecting from the sides of the body.

Saichania, the best known and longest (seven metres) of the three, is known from particularly well-preserved remains and in addition to the intricate cobblestone-like dorsal spines, also had a spike-studded bony collar protecting the neck. This genus also shows considerable ventral armour under the belly, while the somewhat flattened skeleton is massively constructed.

Talarurus, the oldest genus shown, was also a very large animal reaching some six metres in length, and having a very deep, barrel-like body similar to a hippopotamus. The detail of the armour is more speculative than that of *Saichania* but is believed to have been similar to the banded arrangement of long, bony spines shown here. All three animals were, like all ankylosaurid genera (as opposed to nodosaurid ankylosaurs), equipped with heavy, bony tail clubs formed of ossified bones and operated by powerful sets of muscles connected to the hindlimbs and the pelvic area. Although slow moving by nature, fossilised trackways of the closely related nodosaurs show that ankylosaurids walked with the limbs held under the body rather than splayed out to the sides. Like crocodiles, ankylosaurs were quite capable of greater speeds when the situation demanded.

Food was primarily sought for using the well-developed olfactory senses which included the Jacobson's organs (pouch-like olfactory structures connecting the nose with the mouth that are particularly well developed in snakes) as well as a long and highly mobile tongue. While low, shrub-level vegetation was the main constituent of the diet, ankylosaurs had provision for taking maximum advantage of the mid-Mesozoic flourishing of insects that accompanied the evolution of the flowering plants. Ankylosaur diets undoubtedly included a wide range of invertebrates such as termites, beetles and grasshoppers as well as insect larvae. The mobility of the tongue during searching for such prey was supplemented by the high degree of vertical and particularly horizontal mobility of the head, as evidenced by the shape of the occipital condyle which connects the skull to the neck vertebrae.

The Gobian ankylosaurs show greater diversification than the two North American genera *Ankylosaurs* and *Euoplocephalus* but appear to be less advanced in some aspects. Conversely, they also exhibit characteristics unknown in their western relatives, for example, the completely ventral orientation of the occipital condyle and the very strong ventral armoury. As the Gobian species evolved, they showed a tendency towards a more massive skull, stronger ossification of the orbit region, and stronger ossification of the palate. Maryanska considers that stegosaurs and ankylosaurs shared a common ancestor (both groups are originally quadrupedal), with relatively unspecialised Asian forms probably migrating across the Bering Straits region to western North America at an early stage of their evolution, after the appearance of *Talarurus* but before that of *Pinacosaurus*.

Opposite: Tarchia *and* Talarurus *(based on G. Olshevsky);* Saichania *(based on W. Skarzynski)*

armour for defence. Glyptodonts had a complete bony carapace over the back and sides of the body, while the tail was encircled with bony rings that in some genera such as *Doedicurus*, ended in a solid club covered in spikes, like a mace. This armour would have been a useful defence against South American carnivores, which included leopard-sized sabre-toothed cats of the genus *Thylacosmilus*.

Armoured glyptodonts of the genus Doedicurus *and a sabre-toothed cat (*Thylacosmilus*) from the South American Pleistocene, about 500,000 years ago.*

4.3. Nemegt, Valley of the Dinosaurs

As Efremov's convoy moved south, the open, panoramic desert flats typical of the area around Ulan Bator began to give way to a more undulating topography of long, valley-like depressions separated by chains of craggy, mountainous massifs. As well as feral horses, Bactrian camels (*Camelus bactrianus*), and rodents like the marmot (*Marmota sibirica*) and the leaping jerboa (*Allactaga bullata*), which inhabit the open desert, these areas in the south were also home to mountain sheep (*Ovis ammon*) and ibexes (*Capra ibex sibirica*). These agile ungulates sported massive, spiral horns as they clambered effortlessly over precipitous cliff faces. Dzheiran gazelles (*Gazella subgutturosa*) were also to be found here, and being capable of speeds greater than 70 kilometres per hour, were easily able to outrun the vehicles over the stony desert terrain.

Not far from Mongolia's isolated southern border with China, the Russian palaeontologists came upon a vast topographic depression, surrounded on all sides by a "sea" of undulating desert. This area, which is known as the Nemegt Valley or simply the Nemegt, is particularly large, stretching for 180 kilometres east to west, and varying from 40 to 70 kilometres north to south. On its northern and southern sides, it is bounded by rugged, low mountain ranges of Palaeozoic origin, the highest peak reaching some 3000 metres, and forming ideal natural borders to isolate the valley from the outside world. Within the mountainous perimeter, the slopes gradually lessen in steepness as they taper off into the sandy basin which forms a huge collection area for Upper Cretaceous and Lower Tertiary deposits.

Similar remote areas in the world's more tropical latitudes, such as the Congo Basin or parts of the Orinoco and Amazon River systems had historically given rise to both scientific and public speculation concerning the question of whether dinosaurs were completely extinct. One such area, in southeastern Venezuela, is the geologically ancient plateau of the Gran Sabana, a mysterious and awesomely beautiful world of undulating grasslands and sheer-walled, flat-topped mountains called "tepuis". The summits of many of these plateaus are so inaccessible and isolated, and their flora and fauna so distinctly different from that of the surrounding grasslands below, that 19th century scientists seriously considered the possibility that dinosaurs

Overlooking the vast Nemegt depression, the "Valley of the Dinosaurs", photographed on the 1946 Reconnaissance Expedition. This area proved to be a huge repository of late Cretaceous dinosaur remains. (From the archives of A.K. Rozhdestvensky, courtesy of S.M. Kurzanov, Russian Palaeontological Institute)

may have survived upon them. The most famous tepui, Roraima, was finally conquered in 1884 by an expedition financed by the Royal Geographic Society, the British Association, and the Royal Society. Many new species of small animals have been found on these isolated blocks of landscape, and the fact that nothing even remotely resembling a pterodactyl was found on the 64 square kilometre boulder-strewn summit of Roraima did not dispel the public's speculation. Sir Arthur Conan Doyle's 1912 novel *The Lost World* perpetuated the concept of surviving dinosaurs on an isolated plateau in such a convincing fashion that one American university actually organised an expedition to South America to locate the area. The Nemegt Valley would prove to be a fitting Central Asian analogy to Conan Doyle's surrealistic world, being just as well if not more densely populated with dinosaurs, albeit in a somewhat more inanimate state at the time of their discovery.

A haunting "Lost World" atmosphere was certainly evident to Efremov's group as they travelled over the open flats of the Nemegt Valley. Constant winds carried out a relentless cycle of picking up fine-grained deposits from the outlying slopes and scattering them over the valley floor where they accumulated into gigantic sand dunes, sometimes kilometres in length. Should an armoured ankylosaur have suddenly appeared from behind one of the dunes, sand falling from its spine-studded back, it would not have looked out of place in such an area. Although the environment was semi-arid, the valley was not without significant vegetation and in some areas scattered bushes of the small scrubby shrub *Haloxylon* dotted the landscape for as far as the eye could see. The stunted branches of this common plant are stored by Mongols in summer and used as winter fuel. Other less prolific plants included thorny caragana bushes, wild garlic, and hardy shrubs of the genus *Nitraria*, bearing sour-tasting berries. Rainfall of course, was a rare event, and even when it did occur, the hot air rising from the valley floor was so intense that it often evaporated before it reached the ground. But precipitation was heavier over the mountainous peaks around the perimeter. There, the waters caused high levels of erosion, cutting deeply into the sandy deposits at the base of the slopes and resulting in seemingly endless, sometimes vertically-sided ravines. Rain water collected in such areas, forming small channels which eventually, by circuitous routes, found their way down to the valley's topographic centre where they emptied into groups of shallow brackish lakes.

A landscape of confusing dunes within the Nemegt Valley, photographed in 1946. (From the archives of A.K. Rozhdestvensky, courtesy of S.M. Kurzanov, Russian Palaeontological Institute)

The palaeontologists uncovered many dinosaur cemeteries scattered across 100 kilometres of the Nemegt Valley, and these contained the remains of many Cretaceous dinosaur types including hadrosaurs, carnosaurs, ankylosaurs, sauropods, and ornithomimosaurs, among others. A complete review of all of the finds made here is beyond the scope of this book, but several excavations in particular proved exceptionally rewarding. A sequence of hard, fossiliferous sandstone outcrops at the locations "Nemegt" and "Altan Ula" at the northern sector and "Tsagan Khushu" further to the south were the most important sites, with the work here being carried out under waterless, harsh conditions, the nearest wells being some forty kilometres distant.

On May 9, (celebrated as World War II victory day in the Soviet Union), Rozhdestvensky and the others held their own celebrations after one of the members had located the complete skeleton of a giant predatory carnosaur, which was immediately christened "Eaglon's skeleton" after its finder. It was discovered among an inhospitable series of gorges composed of heavily dipping red sandstone strata, the summits of which were so eroded that Rozhdestvensky describes them as resembling the shapes of towers, spires, and sphinxes. On the summit of a hill was where the great ten-metre-long skeleton lay, on its side, with head thrown back (an attitude frequently encountered in camels that have perished in the desert, and is due to contraction of the cervical tendons, not death by thirst as has often been assumed). Eventually, six more complete skeletons of this formidable reptile, designated as *Tarbosaurus bataar* (=alarming reptile from Bataar) were retrieved on this series of expeditions, and at least eight more have subsequently been found.

Below: *Tyres don't last long in the stony desert terrain. Head taxonomist A.K. Rozhdestvensky seated at left, photographed during the 1949 season.*

Right: *The 1948 Expedition camp at the locality "Nemegt" at the northern sector of the Nemegt Valley.*

Below right: *1948 Expedition members at the Altan Ula excavation site (left to right) J. Eaglon, A.K. Rozhdestvensky, M. Lookijnova.*
(All photographs from the archives of A.K. Rozhdestvensky, courtesy of S.M. Kurzanov, Russian Palaeontological Institute)

Tarbosaurs were undoubtedly common in this part of Central Asia during the late Cretaceous. This is unusual, because in any community of animals the carnivores at the top end of the food chain are heavily outnumbered by the herbivores, since it takes many herbivores to feed each carnivore. Even by dinosaur standards, tarbosaurs were large animals, and some adult specimens found measure fourteen metres in length. Such a specimen would have stood between five and seven metres high when walking on the strong, hind legs. *Tarbosaurus* appears to be most closely related to the genus *Tyrannosaurus* from North America though a full grown *Tarbosaurus* was longer and more slenderly built, with a more gracile and slender skull.

Because of their size, it is natural to assume that tarbosaurs were the predators that pitted themselves against the stubborn ankylosaurs, using the giant jaws and clawed feet in the offence, though they may also have preyed on much larger dinosaurs as well as scavenging the carcasses of animals that had died naturally.

The excellent series of well-preserved tarbosaur skeletons now known from the Nemegt represents a range of animals of different ages from young juveniles to old adults, and as is the case with *Protoceratops*, these present palaeontologists with one of the rare opportunities to study the various growth stages of a dinosaur species.

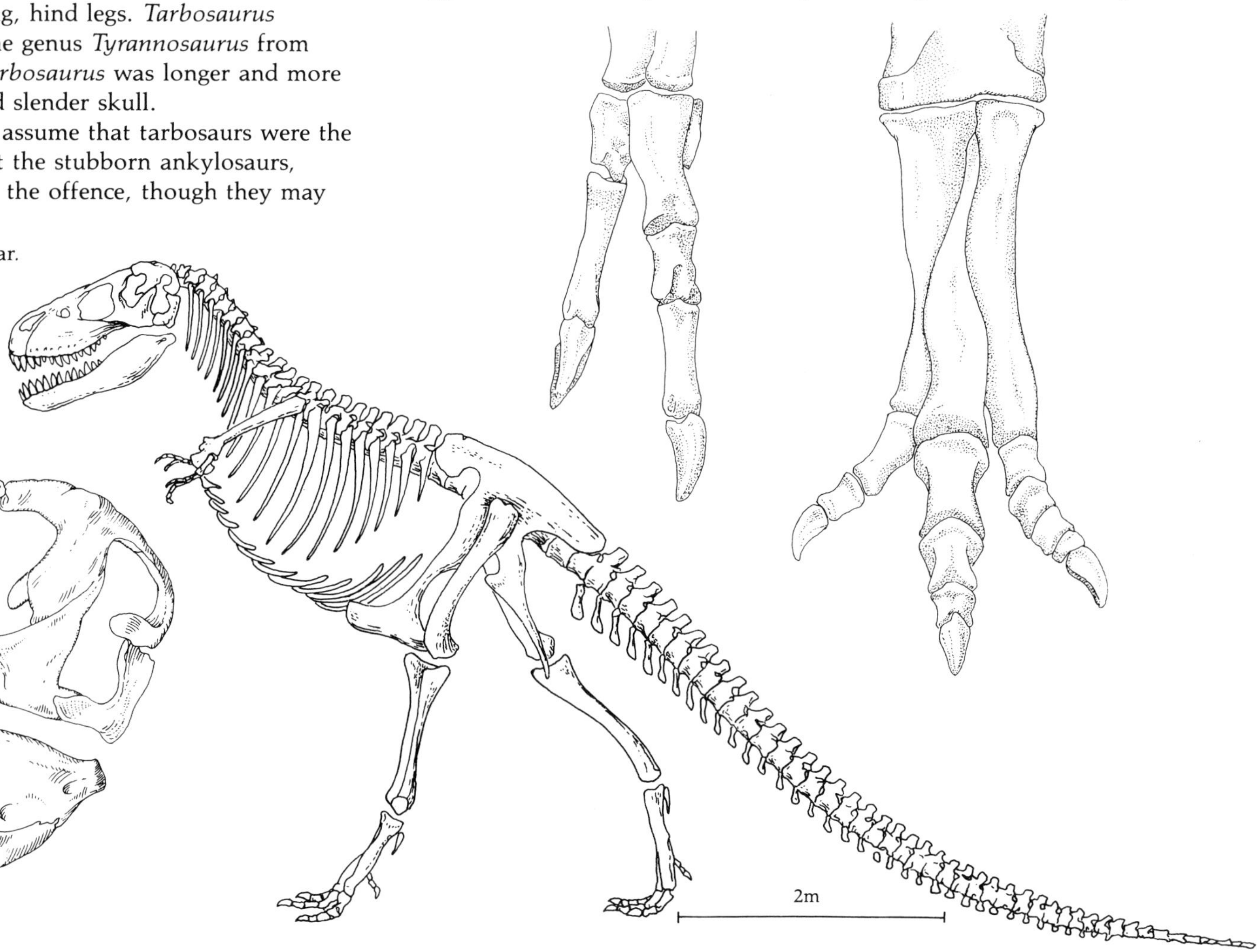

Reconstructed skeleton of Tarbosaurus bataar.
Below: *Skull of* T. bataar.
Right: *Hand and foot of* T. bataar.
(Based on Maleev)

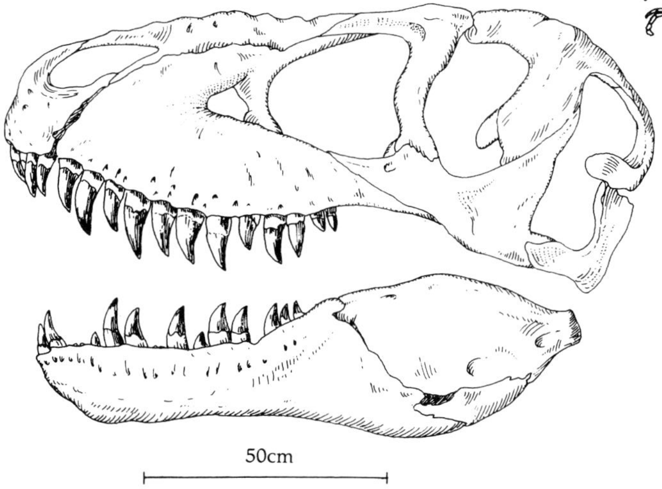

As tarbosaurs grew, in addition to the body naturally becoming more robust and powerful, the shank and foot became relatively shorter compared to the body, the brain crest became enlarged, and the snout lengthened. Unlike some other large theropods though, the teeth appear to have remained the same size relative to the skull throughout life. It is believed that carnosaurs like tarbosaurs were capable of attaining an age of some sixty years, if not considerably more.

Popular texts (not to mention films) have often made much of the assumed titanic struggles between carnosaurs and their prey (in particular the North American horned ceratopians such as *Triceratops*). But how much food would a tarbosaur have actually required, and how often would it have needed to hunt or scavenge? Estimates of the energy for a reptile of this size indicate, for example, that a six tonne animal would have had a daily requirement of some seventy kilograms of meat. The great strength of tarbosaur skulls allowed them to consume probably over 80 percent of any given large carcass, so if a tarbosaur fed on a large herbivorous dinosaur such as a hadrosaur of its own size or larger, it would have gorged itself for days until it was completely satiated, gaining sufficient energy for well over a month of inactive life. Of course, how much of the carcass was consumed would also depend on the number of other tarbosaurs or smaller theropods that might share in the meal.

With respect to the frequency of feeding, tarbosaurs were not unlike modern constrictor snakes. The elastic jaw mechanisms of anacondas and large pythons enable them to devour entire animals the size of

An African rock python (Python sebae) *devouring a Thomson's gazelle.*
(Photo by G. Ziesler, courtesy of Bruce Coleman Picture Library)

which is difficult to believe unless actually seen, after which they become torpid for many weeks while the giant meals are digested. Although some, like the boa constrictor, are capable of digesting a large meal in seven to nine days, the speed of digestion varies and is greatly reduced with a lowering of temperature, often stopping for long periods without any harm to the snake. (Pythons kept in captivity have gone for as long as thirty months between large meals.)

Therefore, if we were able to return to the steppes of Central Asia in Cretaceous times, rather than witnessing mortal combat between giant carnosaurs and equally giant herbivores at every turn, we would instead observe small groups of tarbosaurs of mixed ages and sizes, lying about resting in the shaded seclusion of forest clearings or groves of trees, or at the edges of the expansive but shallow lakes that flooded large areas of this part of Asia. Within and around these same lakes would be herds of herbivorous dinosaurs such as sauropods and hadrosaurs, passively feeding in apparent indifference to the great carnivores lying near them. An analogous scene involving well-fed lions and herds of zebras or wildebeest can today be observed in many parts of Africa.

Given that hunting did not take a demanding amount of time, can we make any meaningful assumptions about other aspects of tarbosaur behaviour, such as reproduction and care of the young? Once again, we look to modern large reptiles as well as birds for comparisons, and in particular, the close archosaurian relatives of the dinosaurs, the crocodiles. Historically much maligned by humans, recent observation has shown that these reptiles exhibit an unexpectedly complex parental behaviour that gives us some indication of possible comparable behaviour in carnivorous dinosaurs.

Male Nile crocodiles, for example, establish and maintain fixed breeding territories, fiercely defending them against any male intruder. Mating with a receptive female takes place following a highly ritualized courtship display in the water. When the female is ready to lay, she proceeds to dig a large depression in an area of dry sand. The nest may be up to five metres in diameter and is often situated among many similar nests within communal nesting grounds that are used repeatedly season after season. During the night, between forty and eighty eggs are laid in several tiers or layers, each layer being dutifully covered by sand until the nest is completely filled. These nests may have an internal temperature variation of as little as 3°C.

Tarbosaurus bataar

Unearthed from Upper Cretaceous strata in the Nemegt Valley, *Tarbosaurus bataar* (= alarming reptile from Bataar) is one of the few dinosaur species for which complete skeletons from several different age groups is known. These include seven (three from the "Dragons' Tomb" site) collected by the Russian Expeditions (1946–1949) and at least six more collected by the joint Polish-Mongolian Expeditions between 1963 and 1971. First described by Maleev in 1955, *Tarbosaurus* is the longest of the carnivorous dinosaurs (fourteen metres) and is closely related to the North American genus *Tyrannosaurus* (12.3 metres), *Tarbosaurus* being more slender, and having a longer and more gracefully constructed skull. The osteology of the foot also differs between the two genera.

The *Tarbosaurus* skull, up to 1.3 metres long, is high and laterally compressed, and characterised by small, rugged bony protuberances on the snout. The jaws are gracile but powerfully built, and armed with a battery of fifteen-centimetre-long backward-pointing teeth. Muscle scars on the jaws show that, as in crocodiles, the muscles opening them were not particularly large, but those that closed them were immensely powerful. Like other tyrannosaurs, the forelimbs are tiny for the size of the animal, the two-fingered hands being equipped with stout claws like grappling hooks. These are often described as vestigial, but they may have been used during mating and nest-building and were certainly used to anchor the body to stop it from sliding forward when the animal raised itself from the ground after resting.

Tarbosaurs walked in a stooping fashion on strong, pillar-like legs, with the long, tapered tail acting as a counterbalance against the skull and body. The foot was powerfully built with three, splayed, clawed digits, and an additional small, non-functioning one at the back that did not reach the ground. The elongated ankle bones suggest that tarbosaurs were capable of greater speeds than other tyrannosaurs, whilst the metatarsal bones were locked together below the ankle for additional strength.

Potential prey of tarbosaurs in the late Cretaceous of Asia included a diverse range of hadrosaurs, iguanodontids, sauropods, and ankylosaurs, though it is not known whether they were particularly active predators or whether they lived primarily by scavenging. Ankylosaurs would have been likely prey of tarbosaurs, and as their formidable dorsal armoury was difficult to pierce, the method of attack would have involved trying to overturn the heavy animals using the jaws and the clawed feet. Against a large ankylosaur, this could prove a dangerous operation, with an inexperienced tarbosaur risking serious injury if toppled or hit directly by the ankylosaur's tail club.

Feeding involved tearing at the carcass with powerful tugging and twisting of the jaws by the strong neck muscles, in much the same manner as is employed by crocodiles. Large portions of food were able to be swallowed entire, this being facilitated by the somewhat flexible construction of the lower jaw. Because of their size and slow metabolic rate, large meals would have provided tarbosaurs with sufficient energy for many weeks of relative inactivity.

Details of many aspects of these animals' behaviour are at present not well known. Much of their lives, which may have been over sixty years in duration, was spent reclining in the manner shown for the reconstruction opposite. As is the case with today's lions of the African savannah, tarbosaurs probably lived in small, social groups comprised of closely related animals of different ages. The remains of nests thought to have belonged to small theropods indicate that animals like tarbosaurs had nest building skills and cared for the young well after they hatched, feeding them with captured food brought to the nests.

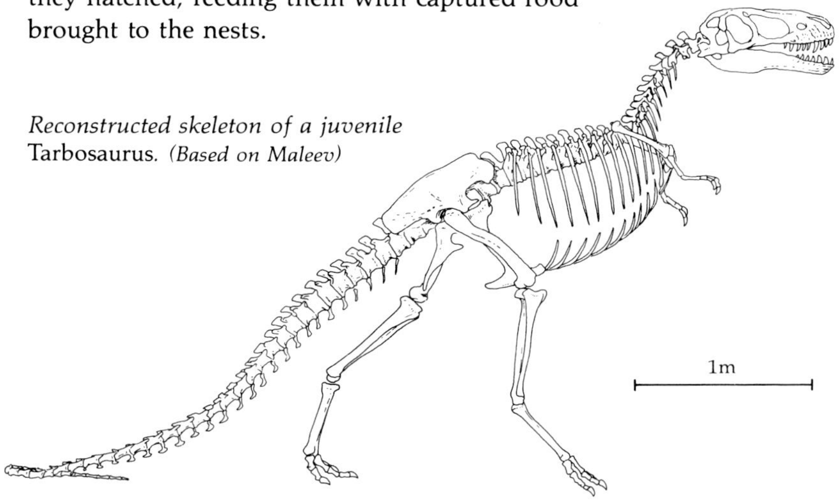

Reconstructed skeleton of a juvenile Tarbosaurus. *(Based on Maleev)*

Opposite: Tarbosaurus bataar, *based partly on J. Sibbick.*

Other crocodilians show variations of such nest-building procedures, such as the American alligator which lays the eggs in a pile of rotting vegetation, turning it at intervals so as to maintain moisture and heat levels within. When the young Nile crocodiles are ready to hatch, they begin to call from inside the nest and the mother assists them in reaching the surface by clawing away the sand. Once all the hatchlings have emerged the female begins an unusual ritual that has remained unchanged for millions of years. Using the powerful jaws capable of completely dismembering large animals, she delicately picks up the hatchlings one by one, and holds them in her mouth. Up to six young at a time are carried in her mouth pouch in the lower jaw, the teeth stopping them from falling out, as she carries them down to a special nursery pool where both she and the father can guard the young as they swim about and learn to hunt small prey. This protective behaviour of the parents is continued until the young are large enough to escape from predators and fend for themselves.

The large number of tarbosaur remains found in the Nemegt area, particularly the number of juvenile animals, strongly suggests that this area contained a tarbosaur nesting site. Although actual nests of this genus have not yet been found, (while the remains of herbivorous dinosaur nests are now quite common, those of carnivorous dinosaurs are virtually unknown), a recent joint Russian-Mongolian Expedition found what appears to be the remains of nests of somewhat smaller carnivorous theropods. Located in the eastern Gobi at the Builjasutuin-Khuduk site, these offer a rare opportunity to study the nesting habits of carnivorous dinosaurs.

The original nests comprised an open, shallow depression on the ground, and probably contained more than one layer of eggs separated by plant material that acted as an incubator. The nest structure appears very similar to those of the modern mound-building birds such as the Micronesian jungle fowl (family Megapodiidae). After hatching, the activity of the young theropods thoroughly mixed the broken eggshell fragments with the vegetation. The very small size of shell fragments indicates that the young, like those of the Nile crocodile, were not precocious but rather, stayed in the nest for a long period. A large number of bone fragments found within the nest appear to be the remains of animals caught by one or other of the parents and brought to the nest to feed the young. None of the remains contain traces of gnawing or biting as would be expected if

In the Nemegt Valley during the late Cretaceous, an adult Tarbosaurus *offers food to her hatchlings within a vegetation-lined ground nest.*

they had been fed upon by the adults. Most of the fragmentary bones belong to the small ceratopian *Psittacosaurus*, a relatively slow moving animal (see previous chapter) as well as assorted bones of other theropods, sauropods, and small scincomorphous lizards (family Paranecellodidae). Kurzanov and Mikhailov who described the nest in 1989 speculated that due to the size of some of the prey captured, the adults must have been theropods considerably smaller and more agile than tarbosaurs. It is reasonable to assume that larger theropods like *Tarbosaurus* showed similar nesting behaviour, which was crocodile-like in some respects (the layered arrangement of eggs and the active young staying in a nursery-type environment for a long time after hatching), as well as bird-like (bringing food for the young to the nest).

What kinds of large herbivorous dinosaurs had shared the well-watered Mongolian landscape with the tarbosaurs and smaller theropods in Cretaceous times? At some sandstone outcrops at Altan Ula, Efremov's group were to come upon just such a find. The Altan Ula outcrops were set among a confusing labyrinth of steeply sloping ravines and hills that extended over many square kilometres. Superficially, the rock faces appeared unfossiliferous, an apparently barren series of deeply cut, shaded gorges of sandstone and eroding, crumbling rock. But one of the expedition's drivers, Pronin, came across a two-metre-thick slab of very hard red sandstone that proved to be anything but unfossiliferous. Embedded in the matrix were the exposed sides of great rib bones of a new species of giant duck-billed hadrosaur. Somewhat reminiscent of the celebrated 1878 European find of many iguanodontid skeletons in Bernissart, the Altan Ula strata were subsequently found to contain the complete skeletons of seven giant hadrosaurs. The great reptiles appeared to have been entombed together, as if in a common grave, and so the locality was christened quite appropriately as the "Dragons' Tomb".

The "dragons" were designated by Rozhdestvensky as *Saurolophus angustirostris* (= crested reptile), and were herbivorous, solid-crested hadrosaurs that most likely led a semi-aquatic existence around and within the shallow but extensive lake and river systems of the Gobi area during the Cretaceous. They were among the largest and most spectacular of the hadrosaurs, reaching lengths of over twelve metres, while one adult skeleton recovered at Nemegt and now mounted in the lobby of the Moscow Palaeontological Museum, towers over 7.7 metres above the floor, more than two metres taller than a bull giraffe. The one-metre-long skull (not including the head crest), was long, high and very narrow, superficially not unlike that of a horse when viewed from the side. As is characteristic of hadrosaurs, the front of the upper jaw terminated in a rounded duck-like bill, while the nasal bone was greatly elongated and extended backwards and upwards about forty centimetres to produce a helmet-like crest at the rear of the skull. The long tail, deep and flattened laterally, was one and a half times as long as the rest of the body and was probably used during swimming. Osteological remains were not the only traces

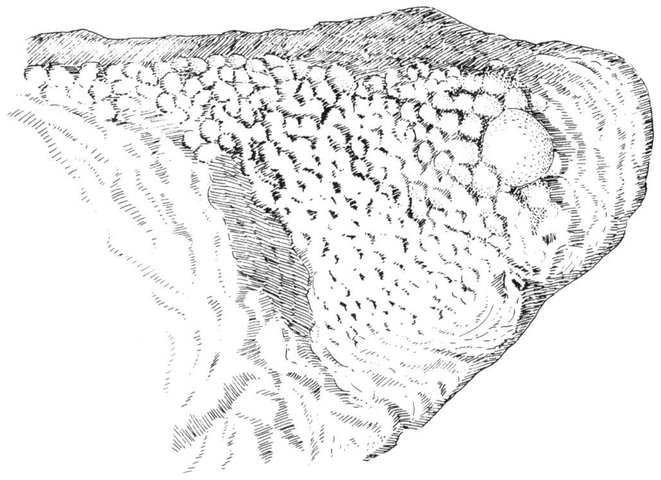

Preserved Saurolophus *skin impressions from the Dragons' Tomb, Nemegt Valley. (Based on Efremov)*

left at the Dragons' Tomb by these great reptiles of over 60 million years age; alongside the seven skeletons and clearly imprinted onto the sandstone surface were the skin patterns of a large dinosaur, most likely that of *Saurolophus*. They show patterns of closely set, bubble-like circular plates pressed into the then soft mud as the giant reptiles had laid down on their flanks upon the shore of a prehistoric Mongolian lake. The impressions must have remained undisturbed and the slow process of petrification would have turned them to stone.

Saurolophus angustirostris

Saurolophus angustirostris (= crested reptile), a solid-crested hadrosaur from Upper Cretaceous strata of the Nemegt Valley, is known from a number of complete skeletons, seven of which were found by a Russian Expedition, buried together at the "Dragons' Tomb" locality. Originally described by Rozhdestvensky in 1952, it is one of the largest hadrosaurs, measuring up to 7.7 metres in height and over twelve metres in length, three-fifths of which is accounted for by the laterally-compressed tail. The distinctive features of *Saurolophus* are the long, narrow skull with the spike-like crest at the back formed from the elongated nasal bones (the ridge of skin behind the crest in the reconstruction opposite is hypothetical). The cartilaginous structure of the nasal cavity was very similar to that of the modern tapir, suggesting that it functioned as an efficient air conditioner and heat regulator during respiration.

Saurolophus was a gregarious animal of aquatic habits that frequented the shores and shallows of lakes and rivers, and is just one of many known hadrosaur genera whose fossil history begins with the earliest hadrosaur, *Bactrosaurus* (= reptile from Bactria) from the lowest Upper Cretaceous strata of Mongolia. By the end of the Cretaceous, the Hadrosauridae had become the most successful family of the ornithopod dinosaurs.

All hadrosaurs were built upon the same body plan, with deep bodies, long tails and strong hind legs with three-toed hoofed feet. The hands had four fingers (two of which ended in small hooves), and from mummified remains, we know that they were webbed, though it cannot be determined if the feet were as well. Like other ornithopods, the toothless horny, duck-like bill of hadrosaurs was suitable for grasping and pulling foliage, while several rows of ridged and tightly packed, self-replacing teeth designed for grinding such fibrous food were situated near the rear of the mouth.

On the basis of skull morphology, hadrosaurs can be divided into three groups; 1) flat-headed forms (without crests) such as *Anatosaurus*, 2) solid-crested forms such as *Saurolophus* and *Tsintaosaurus*, and 3) hollow-crested forms such as *Parasaurolophus*, *Lambeosaurus*, and *Corythosaurus*. Possible functions of the hollow crests of this last group have long been debated. It appears most likely that the internal surfaces of the cavities were lined with soft nasal tissue which greatly increased the animals' olfactory capabilities. Conversely, the crests of solid-crested forms such as *Saurolophus* were most likely used as recognition signals. Because there was such a variety of hadrosaur species as well as large numbers of individuals during the late Cretaceous, these crests would have ensured recognition during feeding forays, nesting activities, or during the mating seasons.

Preserved eye sockets, and traces of large optic nerves indicate that hadrosaurs possessed acute, wide-angle vision, while a preserved columnella from one species suggests excellent hearing ability. We can conclude that these reptiles had very well developed olfactory, optical and auditory senses and were keenly perceptive of all that occurred about them. Perhaps more than any other type of dinosaur, hadrosaurs exhibited complex social behaviours, and they cared for and fed their young in large, communal nesting sites.

Whether they were primarily land or water-dwelling animals cannot be conclusively determined. In the 1920s, the German scientist Krausel studied the preserved remains of stomach contents of one specimen, noting that it contained much fruit and seed debris that was apparently secondarily-derived material ingested while feeding in a peat or marshy habitat. Because this material originated from land plants, some palaeontologists have instead used this as evidence that hadrosaurs fed on land. Even if Krausel was not correct in his interpretation, the premise of a non-aquatic mode of life still does not necessarily follow. For example, the hippopotamus, a possible modern ecological equivalent of the hadrosaur, comes onto land to forage by night, but is water-dwelling by day. Given the adaptations that hadrosaurs possessed, it seems somewhat unlikely that they did not follow aquatic habits for at least part of their lives.

Complete skeletons of *Saurolophus angustirostris* and remains of other Mongolian genera are housed at the Moscow Palaeontological Museum, with some material also housed at the Polish Institute of Palaeobiology. Apart from saurolophine hadrosaurs, genera of lambeosaurine hadrosaurs known from Asia include *Bactrosaurus*, *Barsboldia*, *Taxartosaurus*, *Nipponosaurus*, *Procheneosaurus*, and *Tsintaosaurus*. North America has about the same number of lambeosaurine hadrosaur species as Asia, as well as a smaller species of *Saurolophus*, *S. osborni* from Canada.

Opposite: Saurolophus angustirostris, *based on Flerov and N. Parker.*

Saurolophus was, like all hadrosaurs, an animal of gregarious habits and intricate social behaviour, and there is direct evidence that these dinosaurs followed communal nesting procedures and cared for the young with at least as much dedication as the Nile crocodile. A series of excavations in Montana by J.R. Horner revealed groups of nests belonging to the hadrosaur *Maiasaura* (= good mother reptile) preserved at the aptly named "Egg Mountain" locality. The thickness of the deposits in which the same nests occur — there are other nests of hypsilophodontid dinosaurs here as well — shows that, as with *Protoceratops* and crocodiles, the same localities were utilised for nesting over many seasons.

Each nest is separated from the next closest one by about eight metres (about the length of an adult) and is constructed of a circular mound of earth enclosing a depression two metres in diameter and one metre deep, which the female presumably dug out with her feet. From ten to thirty-five ellipsoidal eggs, each twenty centimetres long, were laid either in a circular arrangement, or what seems more likely, were adjusted to sit like this by the mother using her bill. The nest was then covered, as it is in the ground-nesting megapodiid birds, with an insulating layer of vegetation. Remains of adults, juveniles and hatchlings all found together at this site suggest that the parents tended to and protected the young (which were thirty-five centimetres long on hatching) within the nests, probably feeding them with regurgitated plant material in the same manner as do many species of seabird.

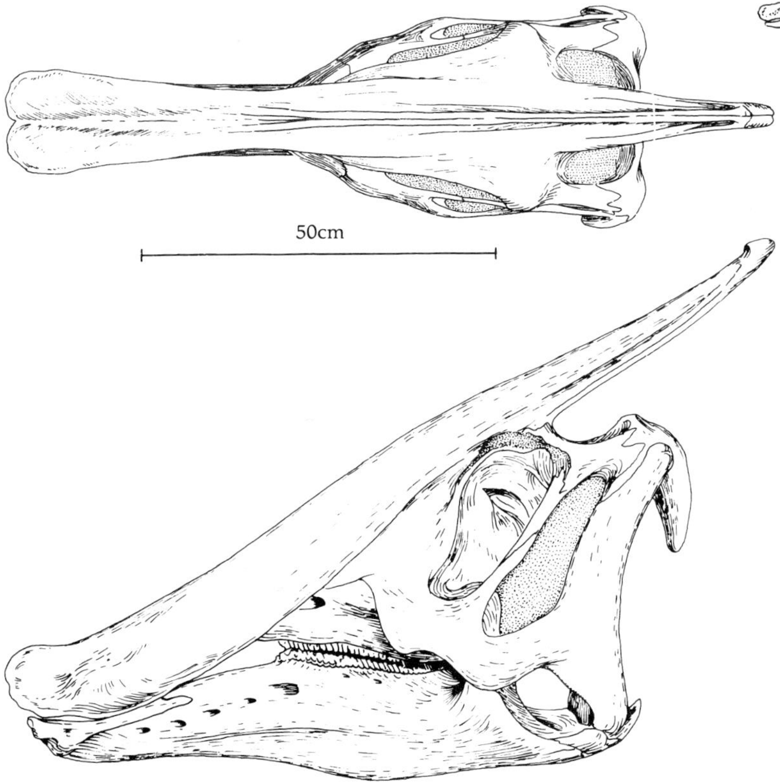

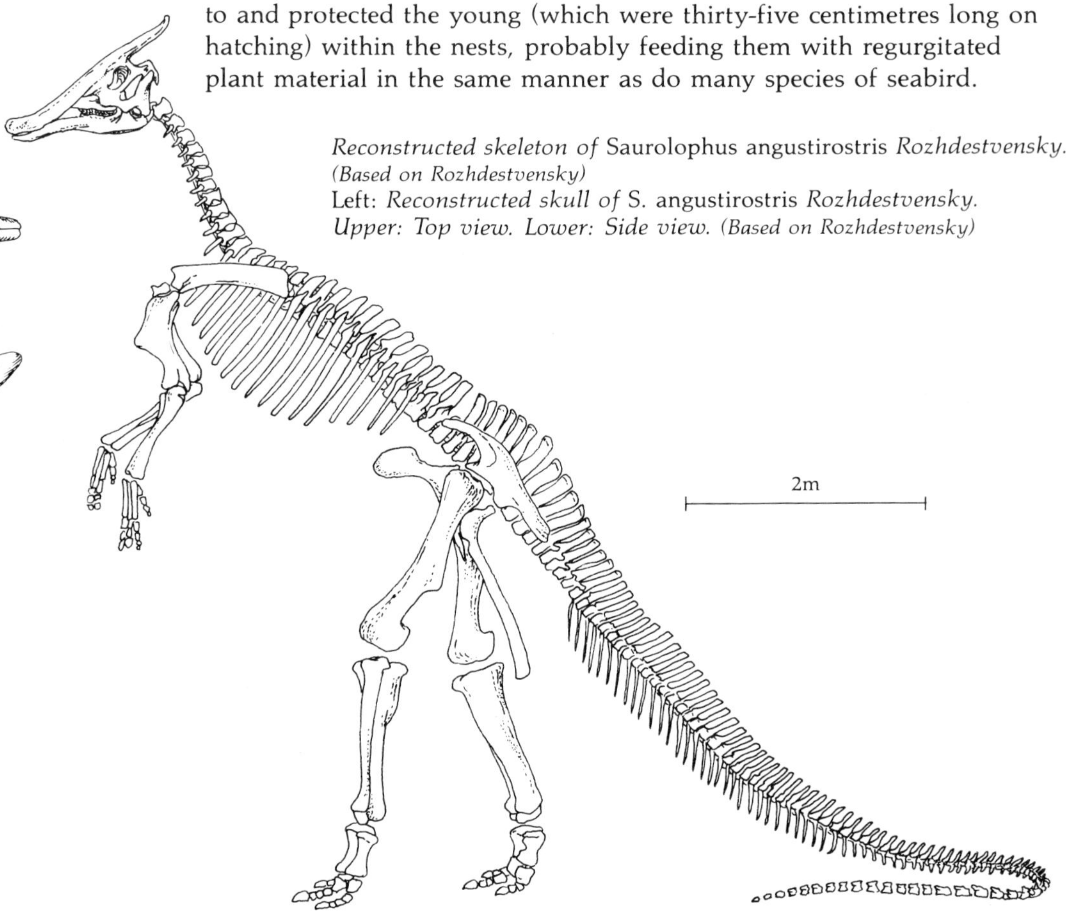

Reconstructed skeleton of Saurolophus angustirostris *Rozhdestvensky.*
(Based on Rozhdestvensky)
Left: *Reconstructed skull of* S. angustirostris *Rozhdestvensky.*
Upper: Top view. Lower: Side view. (Based on Rozhdestvensky)

Excavating at the Dragons' Tomb during the 1948 season. Here, leading technician J. Eaglon prepares a giant skeleton of the hadrosaur Saurolophus angustirostris *for removal from the surrounding sandstone. (From the archives of A.K. Rozhdestvensky, courtesy of S.M. Kurzanov, Russian Palaeontological Institute)*

When the young hadrosaurs reached a length of at least 1.7 metres, they were able to leave the nest site to begin to forage for themselves.

What strange sights these hadrosaur communal nesting colonies must have presented. In an area that provided some shelter from the elements and from predators, as well as being situated not too distant from fresh water and abundant food sources, many huge, crested reptiles carefully moving about between the many nest mounds, each animal selectively tending to the nutritional and other needs of its own group of hatchlings. Apart from the feeding demands of the young, vigilance was constantly required, as there was no shortage, particularly in the Nemegt area, of many kinds of small, lithe and very active carnivorous theropods capable of darting in between the columnar legs of the adult hadrosaurs in an attempt to seize any undefended young.

Above: *Small dinosaur skeletons can be removed and crated for transport at the excavation site. To prevent fragmentation, they are often treated with glues or resins and coated in protective plaster. Much larger skeletons, such as those of tarbosaurs or saurolophs which may be fourteen metres or more long, have to be removed in sections complete with their surrounding matrix of sandstone. Moving these heavy blocks without damage requires much man-power as well as the use of winches and cables. Here, monoliths containing the bones of* Saurolophus angustirostris *are loaded onto a lorry during the 1949 Russian Expedition. (From the archives of A.K. Rozhdestvensky, courtesy of S.M. Kurzanov, Russian Palaeontological Institute)*

Left: *A skeleton of* Tarbosaurus bataar, *one of seven recovered during the 1946-1949 Russian Expeditions, looms over other Mongolian dinosaurs at the Moscow Palaeontological Museum. This particularly well-preserved specimen is from the "Nemegt" excavation site within Nemegt Valley, and shows the deep rib cage, tiny forelimbs, parasagittally-spaced hindlimbs and gracile jaws typical of this species. Skeletons of juveniles are more slenderly-built with relatively larger feet and shorter jaws. Immediately behind is the recently discovered iguanodontid* Probactrosaurus mongoliensis *and a fossilised nest of its eggs. (Courtesy of S.M. Kurzanov, Russian Palaeontological Institute)*

As well as excavating fossils, Efremov and chief geologist Novojilov undertook a geological study of the Nemegt deposits to establish why this area had been so richly endowed with dinosaur remains. The strata here, it appears, consisted of sediments that were deposited at the mouth of an extensive river, now long since dried up. At the Nemegt area, which must have been quite near to the ocean, the river developed into a very large deltaic system, perhaps some forty kilometres wide. Here, the waters deviated to form many smaller channels, lakes and ponds, home to many crocodiles and turtles. The surrounding land supported a prolific growth of forest (large tree trunks, 0.3 metre and more in diameter are frequently found preserved in the Nemegt strata), and subsisting on this abundant food supply were large numbers of herbivorous dinosaurs well adapted to life in and around the delta. These in turn, supplied food, both as prey and as carcasses, for just as many types of carnivorous dinosaurs. With so many animals occupying an environment so suitable for the process of fossilisation, the chances were high that successful preservation would occur. It would not be difficult for the carcass of a dinosaur to become trapped in the soft mud of the delta, where the rates of deposition of deltaic sands and silt were so continuous that the cadaver would soon become covered and later preserved at depth.

In addition to the very valuable ankylosaur, tarbosaur, and sauroloph finds made at Nemegt, there were considerable amounts of fragmentary material collected during the 1948 season. This includes the remains of other identifiable types such as sauropods, ornithomimosaurs, and smaller theropods, as well as some material that is more difficult to adequately classify.

By far the most unusual and enigmatic of this latter material was three giant claw phalanges found in close association with a large metatarsus and pieces of several large ribs. What attracted most attention is the size of the claws — the largest is seventy centimetres

Right: *Skeleton of* Saurolophus angustirostris *from the Altan Ula locality, Nemegt Valley, as exhibited at the Moscow Palaeontological Musuem. Notable features are the elongated nasal bones that form the head crest, the smooth, rounded bill, very robust hindlimbs with splayed, squat digits ending in small hooves, and the laterally compressed tail. In life, the hands were probably webbed while the skin was contoured with irregular bubble-like plates. Behind the exhibits is a section of the mural "Landscape of the Nemegt Age". (Courtesy of S.M. Kurzanov, Russian Palaeontological Institute)*

long (after reconstruction), the smallest (if such an adjective can be used in this case) some thirty centimetres. No claws even approaching these sizes had ever been found. Maleev assigned the claws and associated bones to the same animal, designated as *Therizinosaurus cheloniformis* (*Therizinosaurus* = scythe reptile) the specific name *cheloniformis* reflecting his belief that this animal was turtle-like in morphology (the ribs were somewhat similar to those of the sea-turtle family Protostegidae) and therefore marine in habits. But subsequent

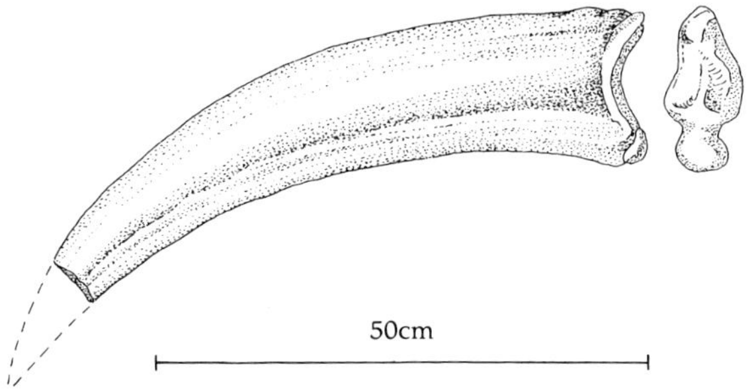

Claw from the hand of Therizinosaurus cheloniformis. *In life, these were covered in a horny sheath, making them longer still. (Based on Rozhdestvensky)*

finds of smaller (though still very large) and similar claws from Kazakhstan and Transbaykalia, in addition to the more recent find of the massive forelimbs and claws of *Deinocheirus mirificus* by the 1965 Polish-Mongolian Expedition to Nemegt (see next chapter), show in fact, that *Therizinosaurus* was a terrestrial, carnivorous dinosaur.

Exactly what were the life habits and appearance of *Therizinosaurus, Deinocheirus,* and related types, may never be adequately known. It does appear that the forelimbs of these animals may have been the strongest part of the skeleton, and hence the only portion durable enough to withstand fossilisation. To hypothesise what these strange reptiles might have looked like, palaeontologists can only compare their remains with what appear to be the most closely related dinosaur types, and use analogies with the habits of animals like extinct and living sloths and anteaters, some of which exhibit similar claw morphology to therizinosaurs and deinocheirids.

4.4. The Beast of Indrik

The Nemegt Valley had proven to be such an unusually rich and rewarding dinosaur-hunting ground for Efremov and his colleagues that it was decided to return and continue to work the locality the following season. For the time being, however, the group travelled northeastwards to reach the area of Lake Orok Noor in the heart of Mongolia. Here, the bones of animals of giant dimensions were once again uncovered, but they were found in Tertiary deposits, not Cretaceous, and were not dinosaur in origin.

Long after the extensive river and lake systems of the Gobi had receded and the dinosaurs, both great and small, had passed into extinction, this part of Asia became the habitat for mammals of unequalled size. During the Oligocene and Miocene periods of the Tertiary, a group of large, hornless rhinoceroses evolved in Central Asia. They were totally unlike any other group of animals, and were to become the largest terrestrial mammals of all time, being exceeded in size only by the larger dinosaurs such as some (but not all) of the sauropods.

The remains of similar animals had been found as early as 1907, with fragmentary bones of the genus *Baluchitherium* (= beast of Baluchistan), having been collected in Pakistan. In 1912, more

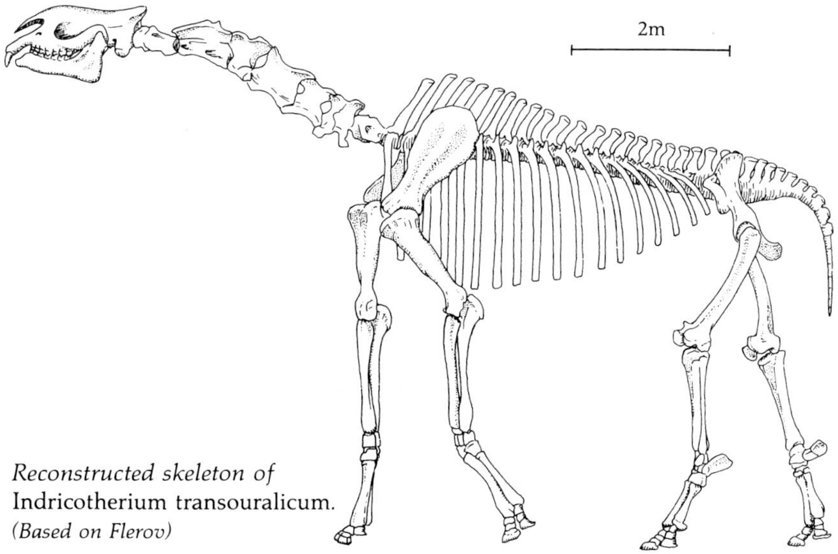

Reconstructed skeleton of Indricotherium transouralicum. *(Based on Flerov)*

complete finds were made in Central Kazakhstan, and in 1915, A.A. Borissyak named this new genus *Indricotherium* (= beast of Indrik, a monster from a 13th century Russian legend 'The tale of the dove'). The Russian Academy of Sciences later conducted extensive excavations in this area up until 1941. The various indricothere remains from the Gobi and other areas were originally classified as many different species by different authors. All of these remains have been studied by the Russian specialist Vera Gromova, who greatly clarified the taxonomy of these animals. She found that *Baluchitherium* was synonymous with *Indricotherium*, with *Indricotherium* taking precedence, and she recognised other valid genera such as *Benaratherium* and *Paraceratherium*. All known species are now placed in the family Indricotheriidae.

The geology of the Kazakhstan area, where the best indricothere remains have been found, is particularly well known. In the early Tertiary, this area comprised large topographic basins of swampy and marshy areas that were heavily forested with familiar species such as walnuts, sequoias and beeches. Around these basins were drier, more

Herds of the giraffe-like Indricotherium transouralicum, *the largest known land-dwelling mammal, roamed the steppes of Central Asia in the early Oligocene. (Based on Z. Burian and Flerov)*

Therizinosaurs and Deinocheirids

Among the most puzzling dinosaur material ever found was collected from the late Cretaceous deposits of the Nemegt Valley in southern Mongolia. These are the remains of giant claws and forelimbs of theropods that have been placed in two families, the Therizinosauridae (= scythe reptiles) and the Deinocheiridae (= terrible-hand reptiles).

The therizinosauridae are represented in Mongolia by *Therizinosaurus cheloniformis*. Giant claws varying from thirty to over sixty centimetres from this animal were found on the 1948 Russian Expedition to Nemegt. Remnants of other possible therizinosaurs (mostly claws), are also known from elsewhere in Central Asia.

The Deinocheiridae, containing the single species *Deinocheirus mirificus*, is represented by two complete trydactyl 2.6 metre-long forelimbs, very similar in morphology to those of ornithomimosaurs, and equipped with thirty-centimetre-long claws. Also found were a shoulder girdle and some isolated ribs. This material was discovered on the 1965 Polish-Mongolian Expedition to Nemegt.

Since these unusual remains were found and assigned, by exclusion, to the Theropoda, some authors have assumed that the animals to which they belonged were giant, carnosaur-like predatory dinosaurs that used their claws in attacking other large dinosaurs. But the known carnosaurs of the Gobi area, the tarbosaurs and related types, tended to have very reduced forelimbs, as well as much larger hindlimbs and large, powerful jaws. If the therizinosaurs and deinocheirids were built upon the same scale and body plan, we are left with an improbably large type of carnivorous dinosaur, whereas the rib sizes of *Deinocheirus* clearly indicate an animal possibly no larger than a tarbosaur. Can we recreate these unusual animals' appearance therefore by considering the functional morphology of their claws and forelimbs and then relating that to their possible taxonomic position within the Theropoda?

Rozhdestvensky has studied the remains of the Nemegt theropods and compared them to other, better known theropods. He suggests that the deinocheirids were descended from the ornithomimosaurs (or possibly another type of small theropod), and that they possessed lightly constructed, toothless skulls. The structure of the wrist and the articulation of the various arm bones allowed the hand to be inturned, a facility found in modern arboreal animals. Therefore, *Deinocheirus*, like the sloth, may have had a torso and hindlimbs that were, in fact, shorter than the forelimbs. Young deinocheirids may also have followed a mode of life not unlike the sloth, moving among the boughs of trees and feeding on fruits and leaves as well as the nestlings of birds or pterosaurs. As they became too large for such activities, they, like the extinct South American ground-sloth *Megatherium*, may have progressed to a life at ground level, pulling down the branches of trees to within reach of the mouth.

In comparison, the laterally-compressed claws of the therizinosaurs, shaped like knife-blades, appear to have been well suited to breaking up ground or the mounds of social insects, such as termites, or stripping the bark from trees in search of other invertebrates. Thus they may have adopted feeding habits analogous to those of the modern anteater, taking advantage of the mid-Mesozoic flourishing of the social insects.

If the deinocheirids and therizinosaurs were the Mesozoic equivalents of modern edentates, they too would have been forest or steppe dwellers, and the scarcity of their remains then becomes apparent, because their habitats were situated some distance from large bodies of water. The most durable parts of the skeleton were indeed the claws themselves, and they were capable of longer preservation than the rest of the body.

Megatherium, *an elephant-sized ground sloth from the recent Pleistocene.* (Based on Z. Burian)

elevated areas of rolling savannah-like grasslands with fewer trees. In the marshy areas, a rich variety of wildlife abounded, including such strange beasts as the chalicothere *Schizotherium*, a horse-like ungulate that possessed large claws instead of hooves, and *Anthracotherium*, a primitive pig-like ungulate, as well as more familiar forest animals such as deer.

But it was in the drier, more elevated regions that the giant indricotheres wandered. In appearance more like colossal horses than rhinoceroses, adult indricotheres weighed about thirty tonnes, stood 5.2 metres at the shoulders, and had deep, cavernous bodies of 8.3 metres length supported on sturdy, column-like legs. The unusual skull, although massively constructed and 1.3 metres in length, was still very small in proportion to the rest of the body, and was supported on a long, muscular and thick neck. Indricotheres are believed to have wandered the Asian steppes in small herds, leisurely seeking food as modern giraffes do, at the upper canopy level of trees. Indricothere finds as far west as Jugoslavia show the extensive geographical range of these animals, but by the end of the first half of the Tertiary, all of these noble beasts had followed their dinosaur predecessors into the realms of extinction.

The palaeontologists continued with their excavations and other field work for as long as the weather permitted, but as the chill Arctic winds begin to bring heavy snowfalls and the night-time temperatures plummeted, Efremov and his group were forced to retreat back to Ulan Bator. During the following year, 1949, a second major expedition was launched, being organised on a similar scale as the previous one and involving some thirty three participants in all. This expedition would retrace the routes of the 1948 one, but would also venture further west than ever before, reaching to the westernmost borders of the country.

Here, at an area known locally as the Valley of the Big Lakes, in the vicinity of Kobdo, a rich fossiliferous deposit of Pliocene mammals

Above: *Before being forced back to Ulan Bator by Arctic winds are Russian Expedition members (left to right) Drs. Rozhdestvensky, Efremov, and Maleev. (By permission of S.M. Kurzanov, Russian Palaeontological Institute)*

Right: *Much of the dinosaur material excavated from Nemegt during the Russian expeditions is mounted in the Museum of the Mongolian Geological Institute at Ulan Bator. Dominating the exhibits is this fine skeleton of an adult* Tarbosaurus bataar. *(Courtesy of Novosti Press)*

was located and excavated. A layer, 1.6 metres thick, contained the remains of thousands of animals that Rozhdestvensky surmised had been killed in flash floods. The variety of remains included skulls of the rhino *Chilotherium* (some up to 0.6 metre in length), woolly mastodons, primitive ancestors of the giraffe, spiral-antlered antelopes, pony-sized ancestral horses of the genus *Hipparion*, and many types of rodents. It was evident that the Pliocene fauna of this area must have closely resembled that of the modern savannahs of East Africa. This area has been named "Altan Teli" after the name of a local "somon" which is the Mongolian equivalent of a county. Then finally, Efremov's group returned to continue excavations at sites that had been worked the previous year, principally in the Nemegt Valley, but also at other localities in the southeastern Gobi.

The dog-sized Hyracotherium, *(=*Eohippus, *the "dawn horse"), represented the earliest stage of equine evolution that occurred in North America during the early Eocene. Its ancestors later migrated to the Old World where by the Pliocene, forms such as the pony-sized* Hipparion *roamed the steppes of the Gobi. Although horses were eventually to die out in America, they became domesticated in Asia. (Based on C. Knight)*

Thus the third and final of this series of Russian Expeditions to the Gobi had ended, and after three seasons of intensive field work, an unequalled stock of Cretaceous Asian dinosaurs had been collected. This included many totally new types as well as good collections of more familiar species already known from previous expeditions. In addition, a valuable collection of Tertiary mammal remains had been made. The cost, in terms of labour, time and investment had been high, but perhaps some concept of the success of these ventures can be gauged from the fact that something like 460 separate wooden crates had to be constructed to transport the monoliths (that is, the bones together with surrounding rock, all coated with plaster for protection), which added to an accumulated weight of over 120 tonnes. A large portion of this total was material recovered from the Nemegt Valley, including no less than ten excellently preserved complete skeletons of tarbosaurs and duck-billed saurolophs. The Nemegt Valley had been established as one of, if not the most important locality for late Cretaceous dinosaurs, and it was to Nemegt that in future decades, palaeontologists from Russia, Poland, and Mongolia itself would return time and time again.

Some of the material collected on the 1946-1949 Expeditions was delivered to the Mongolian authorities, including one of the giant tarbosaur skeletons, later to form part of a permanent exhibit at the Municipal Museum in Ulan Bator. The vast bulk, however, was naturally transported back to the Russian capital where the most time-consuming and scientifically rewarding tasks of all were to begin: the detailed examinations, measurements, comparisons with existing material, and finally, the reconstructions.

Many of the specimens were prepared for permanent display for the general public at the Academy of Sciences' Palaeontological Museum in Moscow. Today this display includes a wide selection of complete skeletons and supplementary material recovered from Mongolia, and offers the public a hauntingly lucid glimpse into the past, at some of the most remarkable animals to have ever existed upon the Asian continent.

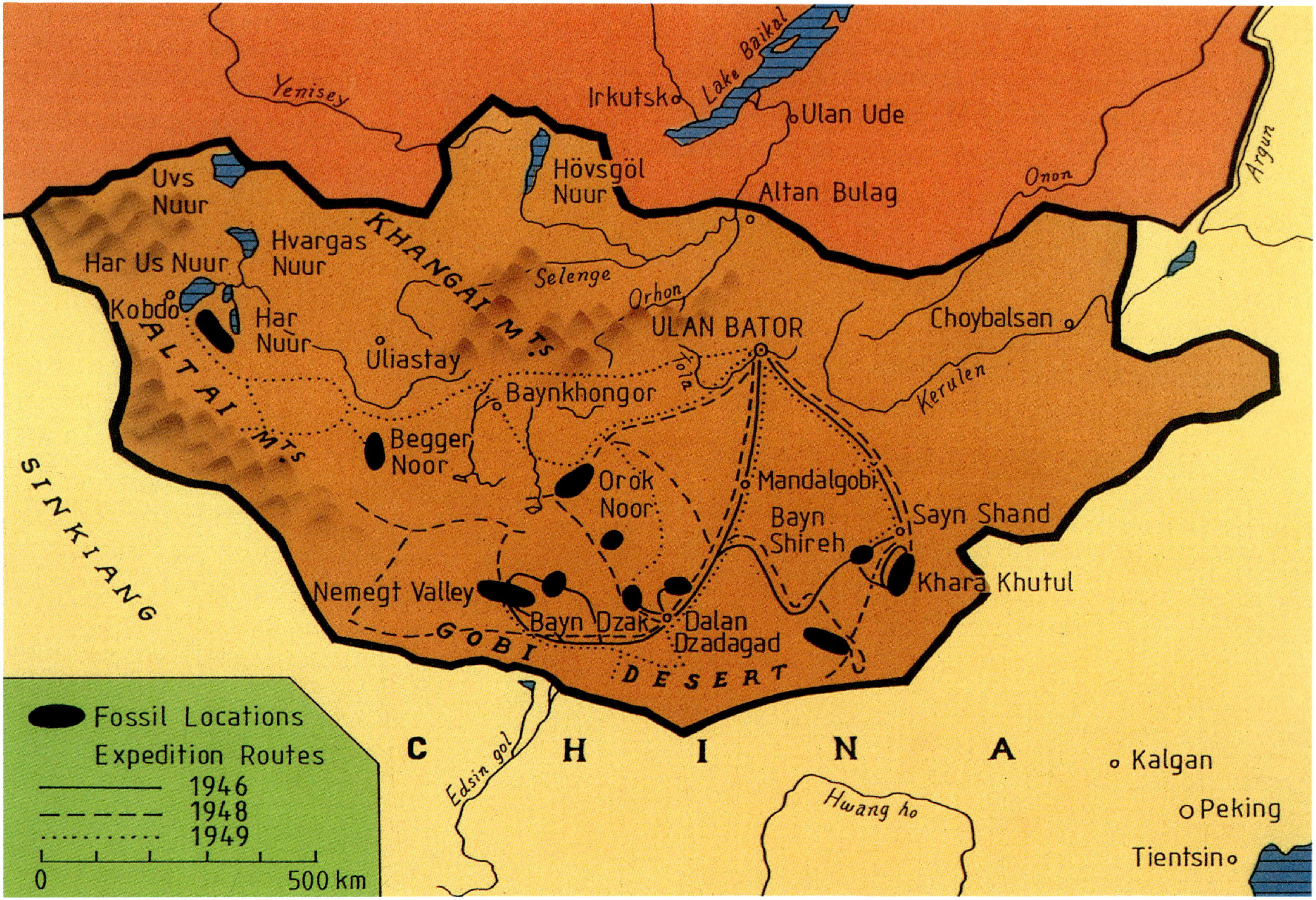

Map of Mongolia showing areas explored by the 1946–1949 Russian Expeditions. (Based on Rozhdestvensky 1957, colour by Jos Coufreur)

Chapter Five

Hunting for Dinosaurs

5.1. The Asiatic Dinosaur Rush

IN 1955 a young palaeontologist who was senior scientist at the Polish Institute of Palaeozoology (part of the Polish Academy of Sciences) journeyed to Moscow where she visited the Museum of the Russian Palaeontological Institute. There, Zofia Kielan viewed with interest the museum's collection of tarbosaurs, saurolophs, ankylosaurs, and other dinosaurs that had been recovered from the Nemegt Valley during the Russian Expeditions of the previous decade. She also met the expeditions' chief taxonomist Anatole Rozhdestvensky, who recounted to her the many adventures he and his companions experienced in the deserts of Mongolia. Zofia Kielan (later Kielan-Jaworowska) had earned her doctorate degree in zoology and palaeontology at the University of Warsaw under the guidance of the then Dean of Polish palaeontologists, Prof. Roman Kozlowski. His series of lectures on the fossil Cretaceous mammals of the Gobi always captivated his student audience, and although Zofia Kielan began postgraduate research on trilobites and Palaeozoic polychaetes, she maintained a keen interest in the Gobi and its fossil vertebrate fauna.

In March 1961, the convention of representatives from the various Academies of Sciences of the COMECON countries of Eastern Europe was held in Warsaw. Present at the convention was the chairman of the recently established Academy of Sciences of the Mongolian People's Republic, Prof. Shrendyb. During the convention, Roman Kozlowski raised the possibility of further exploitation of the Gobi vertebrate fossils, a suggestion that received a favourable response from the other members. A year later, after a fact-finding trip to Ulan Bator by a delegation from the Polish Academy of Sciences, an agreement was signed to organise an initial three year series of expeditions that would involve both Polish and Mongolian palaeontologists. Zofia Kielan-Jaworowska was nominated to lead the Polish side, while Naydin Dovchin was chosen to lead the Mongolian side.

The choice of Polish palaeontologists for such ventures was interesting, because at that time virtually all the research experience of the Polish Institute of Palaeozoology (now the Institute of Palaeobiology) lay in the field of invertebrates, with the only vertebrate fossils studied being Devonian and Cenozoic fishes, as well as some recent mammals from the Pliocene and Quaternary. In fact no museum in Poland possessed a single dinosaur skeleton or bone, or for that matter even a cast of a bone, as that part of Europe where Poland now lies had been submerged during the Cretaceous. Moreover, the Institute of Palaeozoology was totally destroyed (as was much of Warsaw) during World War II. Like many other students at that time, Kielan-Jaworowska had attended lectures in Prof. Kozlowski's private flat, there being no other suitable venue. So even though the operations for these expeditions were taking place in the 1960s with a good knowledge of the previous American and Russian field work, the prospect of successfully retrieving dinosaur-sized skeletons from the Gobi presented just as daunting a task to the Polish palaeontologists as it had to Efremov and his colleagues some two decades earlier.

Nonetheless, with Roman Kozlowski the forthcoming expeditions had the backing of one of the world's leading and most respected palaeontologists, and many of his co-workers likewise enjoyed international reputations. Of course, by that time the Mongolian capital was served by rail links (the Irkutsk – Ulan Bator – Peking line) and airlines to surrounding countries, so all heavy equipment and supplies could be crated and rail freighted to Ulan Bator, with the Polish scientists later flying to the capital from Warsaw. By May 1963, following some eight months of planning and organisation, a small reconnaissance party of eleven Mongolian and Polish scientists entered the Gobi. Their aim was to examine previously located fossil deposits that had been worked by either the Americans or the Russians, including areas in the south and southeastern sectors of Mongolia such as Bayn Shireh, Bayn Dzak, the Nemegt Valley, and some nine localities south of Sayn Shand. Also, new outcrops of Cretaceous and Cenozoic fossils would be searched for and if necessary, fixed for later excavations. With the experience of that summer's field work behind them, the palaeontologists would be adequately prepared for the 1964 and 1965 summer seasons.

By June 1964, the first of the non-reconnaissance joint Polish-Mongolian expeditions had left Ulan Bator to begin field work in the south of Mongolia. So began the third phase of what has been dubbed "The Asiatic Dinosaur Rush". During the 1940s, the Russians had found the value of using heavy-duty trucks working in cooperation with small scout cars capable of speedily covering great distances for daytime observation work. So the Polish-Mongolian expeditions were to be supplied with several Polish-built multi-wheel driven Star 66 trucks loaned from the Star truck factory at Starachowice. These vehicles were used throughout the expeditions and proved ideal for the loose, sandy terrain that characterises much of the Gobi. Other trucks ranging in size from 2.5 to eight tonnes such as Czech Tatras and Russian ZIL and Gaz models would be borrowed when required for short periods from Ulan Bator. These vehicles were used to transport equipment to and from the fossil sites, and to transport the fossil-containing monoliths to the capital. Like the Russian operations, fast Gaz 69 field cars would be used for the scouting and observation work.

As with all expeditions to remote areas, these ventures had to rely to a large extent on manual equipment and much hard physical labour if they were to succeed. The range of tools used in one season (1965) for example, included pickaxes and picks (twenty of each), thirty-three spades, forty hammers, 100 chisels, three pulleys (two mechanical ones fixed to trucks and a hand-operated one), and an array of carpentry and locksmith tools. The packing and protection of the skeletal remains required over two tonnes of plaster of Paris, 150 square metres of canvas, 120 litres of liquid polystyrene, 100 pre-built crates, five cubic metres of timber for building crates on site, and large quantities of wood shavings, cellulose paper and corrugated cardboard. In 1965, 14,000 litres of petrol were shipped from Poland for the three Polish trucks alone. The European-type tents that had been used in 1963 soon proved incapable of withstanding the fierce Gobi winds, so later expeditions would use tents made to Mongolian "yurt" specifications. These were floor-less and of sail canvas construction, being supported by two vertical aluminium poles connected by a duralumin rod.

Expedition members E. Rachtan and T. Maryanska try to secure equipment at the Altan Ula camp during a sandstorm in 1965. (Courtesy of R. Gradzinski, Polish Geological Institute)

Apart from Kielan-Jaworowska, who had since been appointed Director of the Institute of Palaeozoology, other Polish palaeontologists on the 1964 and 1965 seasons included Kazimierz Kowalski (1964), Andrzej Sulimski (1964), Magdalena Borsuk-Bialynicka (1964), Teresa Maryanska (1964, 1965), Halszka Osmolska (1965), Aleksander Nowinski (1965), Josef Kazmierczak (1965), Jerzy Malecki (1965) and Henryk Kubiak (1965). Polish geologists Ryszard Gradzinski and Jerzy Lefeld participated on both expeditions. Palaeontologists on the Mongolian side included Demberlyin Dashzeveg (1964, 1965), Naydin Dovchin (1964, 1965), and Rinchen Barsbold (1965), while extra technical assistants and drivers boosted the totals to sixteen members in 1964 and to twenty-three in 1965.

In addition to working the fossil localities visited during 1963, the 1964 and 1965 expeditions also worked many Cenozoic outcrops further toward the central and western sectors of Mongolia. At Altan Teli, near Kobdo, far to the west of the country, outcrops originally worked by Efremov's expeditions yielded remains of a variety of Pliocene mammals as well as small reptiles such as tortoises, while some eight localities to the north and south of the eastern ranges of the Altai

Mountains produced remains of a rich Oligocene fauna of mastodons, rhinoceroses, horses, gazelles, and rodents. The Cenozoic mammalian fauna of the Gobi appears to have been at least as diverse as that of recent times, and fully justifies Rozhdestvensky's comparison of the faunal assemblage of Altan Teli to that of the East African grasslands.

As expected, the main emphasis of these expeditions would centre on further exploitation of the rich Upper Cretaceous dinosaur-bearing strata of the Nemegt area. Initially though, one month of each season was assigned to work at Bayn Dzak. Continued excavations at the "Flaming Cliffs" site produced further well-preserved skeletons of *Protoceratops* and complete nests of associated eggs, as well as one nest containing a new type of much smaller nine-centimetre-long egg. Bayn Dzak also yielded an almost complete skeleton of the armoured ankylosaur *Pinacosaurus* (see previous chapter), the limbs and girdle of another ankylosaur genus, and over twenty complete and incomplete skulls (fragmentary jaws and mandibles) from Cretaceous mammals representing both placental insectivores and multituberculates which were an extinct mammalian side-branch. The value of these early mammal collections cannot be overestimated, and Zofia Kielan-Jaworowska was later to specialise in the research of pre-Cenozoic mammals.

Most Mesozoic mammals belonged to extinct orders such as the multituberculates, the symmetrodonts and the triconodonts. The therian mammals (the marsupials and placentals) were contemporaneous with these prototherian mammals during the first half of the Cretaceous, though they did not begin to diversify until the Paleocene, after the extinction of the dinosaurs. At the time of the first Polish-Mongolian Expeditions pitifully few fossils of Mesozoic mammals were known. The sum total consisted of a Triassic skull from China, two from South Africa, the late Cretaceous skulls that had been found by the American Museum's Central Asiatic Expeditions, and some isolated teeth and fragmentary jaws. But by the time the Polish-Mongolian Expeditions concluded, some 150 new skulls of Mesozoic mammals had been added, some even with associated postcranial skeletons. These had all come from two late Cretaceous formations, the Djadokhta and the somewhat younger Barun Goyot. The collection from Bayn Dzak alone numbered fifty skulls collected over a period of six seasons. Finding such tiny fossils, most of which were only 1.5 to two centimetres long, occupied between four and fifteen people at a time. It was a difficult task involving crawling around on hands and knees (well bandaged for protection!) while peering through a hand lens for several weeks in temperatures reaching 40°C.

5.2. Exploring the Lost World

After the work at Bayn Dzak was completed, most of the remaining 1964 and 1965 seasons were committed to dinosaur collecting at various locations within the Nemegt Valley. Some excellently preserved skeletons of dinosaurs already known from the Russian expeditions were retrieved here, but some finds added new taxonomic groups to an already comprehensive fauna known to have inhabited the area. The result of two seasons' field work was a collection of dinosaur and mammal remains with a combined weight of thirty-five tonnes, the largest portion being the dinosaur material from Nemegt. This included no less than six incomplete skeletons of *Tarbosaurus bataar* representing a range of ages and sizes from four to fourteen metres in length. These finds further confirmed the unusual abundance of these top trophic level carnivores.

It was to be the discoveries of sauropods, ornithomimosaurs, pachycephalosaurs, ankylosaurs, and a totally unexpected new group of large ornithomimosaur-like dinosaurs that were the most notable results of the Polish-Mongolian work at Nemegt. Sauropods and ornithomimosaurs were not totally unexpected in this area; isolated remains of the two groups had been found by the Russians but they were too fragmentary to permit classification to generic level. The sauropods' main period of development had taken place in the Jurassic, with their remains becoming rare by the time Upper Cretaceous strata are encountered. Yet it was in the late Cretaceous beds of the Altan Ula IV locality that an almost complete postcranial sauropod skeleton was uncovered. At a total weight of twelve tonnes, it proved to be the largest and most difficult single find to extract, requiring some four weeks of heavy labour.

This new genus, which was subsequently named *Opisthocoelicaudia* (=posterior cavity tail) and described by Magdalena Borsuk-Bialynicka, has now been placed within the family Camarasauridae, though it does exhibit several features that are peculiar to that family. Evidence of gnawing marks on some of the bones indicate that the giant reptile had been subjected to the predations of carrion feeders, possibly

Sauropods

The Sauropoda contains the giant and spectacular animals that have always epitomised the general public's conception of dinosaurs. Reconstructions of *Apatosaurus, Diplodocus,* and *Brachiosaurus* are well known and need no particular introduction. The more recently described Mongolian sauropods are notable in that they show features that shed light on sauropod habits as well as being late Cretaceous in age, by which time most other sauropods were already extinct.

The first Gobian species to be described in detail was *Opisthocoelicaudia skarzynskii* (=posterior cavity tail) excavated at Altan Ula during the 1965 Polish-Mongolian Expedition to Nemegt. This sauropod was an estimated twenty metres long, though the neck and head were unfortunately not preserved (the reconstruction opposite being partly speculative). The nearest related species appears to be *Camarasaurus* from western North America, but *Opisthocoelicaudia* has a number of skeletal and muscular peculiarities. The straight back was carried almost horizontally as was the tail, while the limbs, particularly the hind, were very robust. These appear to have been powered by muscles attached to the ilium of the pelvis, rather than by the caudi-femoralis which in other dinosaurs attached to the sides of the tail. The tail vertebrae operated with ball-and-socket joints on their posterior rather than anterior surfaces as they did in other sauropods, while the relatively short and heavy tail may have served as a support when the animal reared up on its hindlegs to browse on foliage. A comparison has been drawn between this behaviour and the similar tripodal feeding posture adopted by giant ground sloths such as *Megatherium*. That *Opisthocoelicaudia* also fed on low vegetation is evident from the neural spines which suggest that the neck was habitually lowered.

Nemegtosaurus mongoliensis (=Nemegt reptile) is known from a fifty-five centimetre-long skull and lower jaw found in the same formation and area as *Opisthocoelicaudia*. Placed in the subfamily Dicraeosaurinae, it is compared most closely with *Dicraeosaurus* from Tanzania and with *Diplodocus* from North America. While it might appear tempting to assign the skull to *Opisthocoelicaudia,* a number of previous skull-to-skeleton mismatches perpetuated in North American sauropod skeletons cautions against doing so. The skull of *Nemegtosaurus* is lightly-built with very large orbits and a comparatively small braincase, and unlike the similar-sized *Diplodocus* skull, has maxilla bones that flare outwards at the end of the snout.

A third Gobian sauropod was found by a Russian-Mongolian team in the southeastern Gobi, and being from the Barun Goyot Formation, is the oldest of the three. Known only from an incomplete fifty-eight centimetre-long skull and lower jaw, *Quaesitosaurus orientalis* (=abnormal reptile) is also classified within the Dicraeosaurinae, appearing closely related to *Nemegtosaurus*. As in that genus, the maxillary bones are flared, though the unusual structure of the occipital bone is unlike that of *Nemegtosaurus* or any other sauropod. The skull is high and heavy in construction while the teeth show little wear. The middle ear cavity indicates keen hearing, while the weak jaw musculature and broad scoop-like jaws suggest a diet of soft, possibly aquatic vegetation.

Several sauropod genera have also been found in neighbouring China, among them *Zigongosaurus, Omeisaurus,* and the slender and very distinctive *Mamenchisaurus*, remains of which were originally excavated at Mamenchi in 1957. This late Jurassic diplodocid has a greater than average number of cervical vertebrae supporting an exceptionally long neck; one 22 metre-long example mounted in the Beijing Institute of Vertebrate Palaeontology & Palaeoanthropology Museum having a neck of no less than 10 metres.

Although sauropods were among the most successful dinosaurs and have a history from the early Jurassic to the late Cretaceous, as is the case with hadrosaurs there is little conclusive evidence as to their life habits. They have been variously described as animals whose enormous weight precluded long-term terrestrial activity as well as being unable to inhabit the deep water in which they have often been depicted because of the great pressure this would have exerted on the lungs. Although current opinion favours a terrestrial existence, the fact that the Sauropoda contained a diverse range of animals that should not be subject to generalisations is frequently overlooked. Fossilised trackways show that entire herds of sauropods entered water certainly deep enough to permit swimming, and whilst the teeth of some species suggest tree browsing, the skulls of others do not show tooth wear patterns that would have resulted from a diet of land plants.

Opposite: Opisthocoelicaudia skarzynskii

Polish members of the 1965 expedition at the Altan Ula camp. Left to right: *M. Kuczynski, A. Nowinski, W. Skarzynski, E. Rachtan, D. Walknowski, H. Kubiak, T. Maryanska, J. Malecki, R. Gradzinski, J. Lefeld, H. Osmolska, M. Lepkowski, Z. Kielan-Jaworowska, J. Kazmierczak, W. Sicinski. (Courtesy of W. Skarzynski, Polish Palaeobiological Institute)*

tarbosaurs, while the head and neck had unfortunately become separated and lost before burial. Like other sauropods, the body of *Opisthocoelicaudia* was deep and bulky, the legs heavy and powerfully built, while the back was carried virtually horizontally, or sloping slightly downwards from the shoulders to the hips. The tail appears to have also been carried in a horizontal position, in contrast to those of some other sauropods that were dragged along the ground during walking. That the neck would have been either carried quite low, or habitually lowered (for example, during feeding) is suggested by the structure and length of some of the neural spines, particularly those close to the neck region.

At about the same time that this skeleton was found, a single sauropod skull was unearthed in the same formation nearby, and as tempting as it must have been to assign this find to *Opisthocoelicaudia*, its features showed closer affinity to the subfamily Dicraeosaurinae (family Atlantosauridae) and appeared to be intermediate in morphology between the genera *Dicraeosaurus* from East Africa (Tanzania) and *Diplodocus* from western North America. The most notable feature of this long and laterally compressed skull, which has been placed in a new genus *Nemegtosaurus*, was the completely preserved sclerotic ring surrounding the very large orbit. This structure is made up of flat, overlapping plates which surrounded and preserved the shape of the eye, and was the first complete such structure to be found in a sauropod. The braincase was relatively small, and there were fifty long and narrow peg-like teeth with chisel-shaped crowns, while the jaw articulation suggests the upper teeth did not bite quite opposite the lower.

The habits of sauropods have been the subject of some debate in recent decades. Because of these dinosaurs' vast bulk and weight, most authorities have traditionally viewed them as amphibious animals that rarely ventured ashore, perhaps only for such activities as egg laying.

The second camp of the 1965 Polish-Mongolian Expedition at Altan Ula IV, Nemegt Valley. Just behind the vehicles are the low Cretaceous sandstone hills and ravines with the Altan Ula mountain range in the background. (Courtesy of R. Gradzinski, Polish Geological Institute)

Water would have effectively buoyed up the animals' massive bodies and provided refuge from carnosaur predators (which were assumed to have been incapable of swimming) while it was noted that some genera such as *Brachiosaurus* had the nostrils raised on top of the skull. Animals such as crocodilians and hippopotami have similarly positioned nostrils, and such a feature enables them to breathe while almost completely submerged, with only the eyes and nostrils above water. In addition, the teeth of some sauropods appeared quite weak and were often described as suitable only for coping with soft water plants. However, the teeth of *Diplodocus* and *Brachiosaurus* skulls did show considerable abrasion that suggested a tougher diet than aquatic vegetation. Henry Fairfield Osborn described sauropod vertebrae as hollowed-out and lightly constructed, which is an adaptation for enhanced mobility in large land animals.

By the 1950s, Kenneth Kermack of London's University College had shown that the water pressure on the trachea and lungs of a sauropod standing submerged with its neck erect would be so great that it would have precluded inhalation. For the same reasons, the blood pressure in the neck and head would have had to be unbearably great. Kermack's work was based on the German physiologist Robert Stigler who, in a series of experiments begun in 1911, had proved that human subjects were unable to breathe effectively just one metre below water level. Conversely, Edwin Colbert noted that whales are capable of breathing when they surface, even when the body is inclined with the lungs lying at considerable depth. Whales are able to do this because like other aquatic mammals such as porpoises, they have a protective casing around the pleural cavity to minimise the effects of the water pressure, and Colbert has surmised that sauropods may well have developed similar adaptations.

Since the 1970s, Robert Bakker has argued that many aspects of sauropod anatomy argue strongly in favour of a terrestrial existence, including the pillar-like legs with their elephantine toe and foot structure, the narrow and light rib cage, and the strengthened back vertebrae. In his view, the long neck was most likely used to browse, giraffe-style, at the upper canopy level of cycads and pines. Largely as a result of his work, opinion began to sway in favour of terrestrial habits,

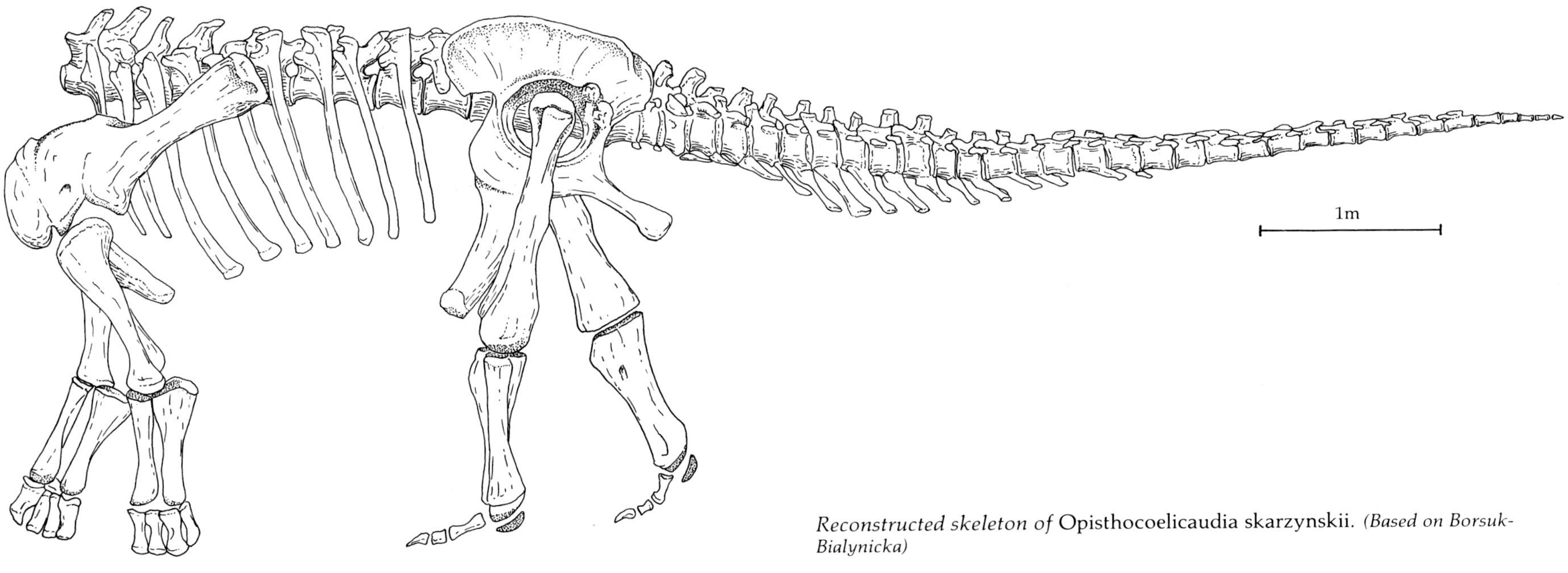

Reconstructed skeleton of Opisthocoelicaudia skarzynskii. *(Based on Borsuk-Bialynicka)*

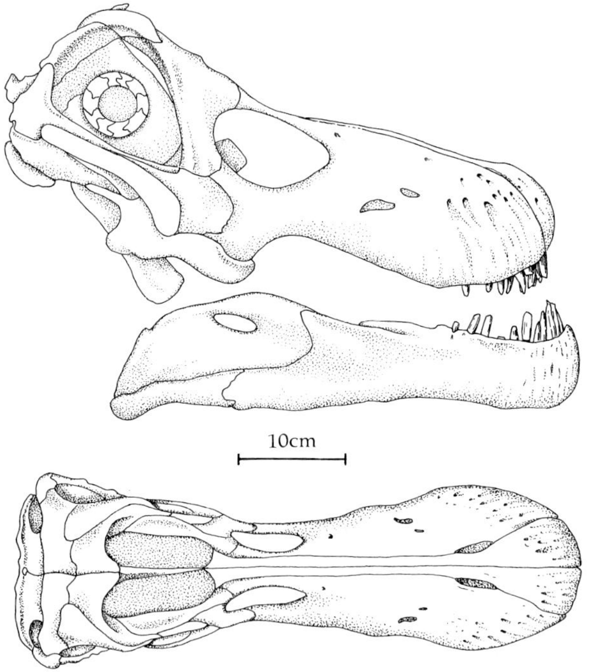

Reconstructed skull of Nemegtosaurus mongoliensis. Upper: *Side view.* Lower: *Top view. (Based on Nowinski)*

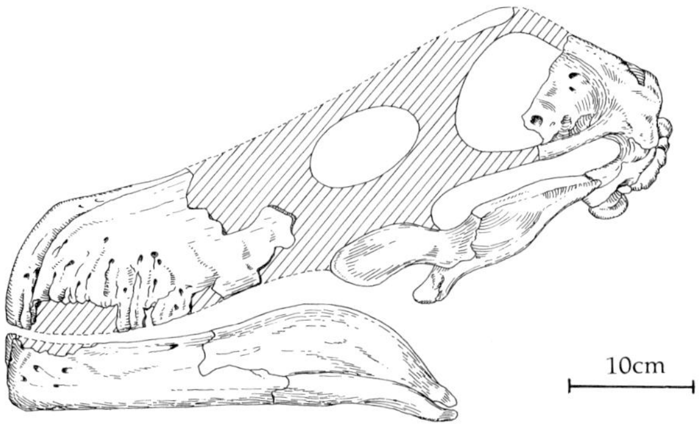

Skull of Quaesitosaurus orientalis, *a Gobian sauropod discovered in the Barun Goyot Formation at Shara Tsav by a Russian-Mongolian expedition. (Based on Kurzanov & Bannikov)*

and sauropods were increasingly depicted as land-dwelling animals that wandered the Mesozoic landscapes much as the elephant today wanders the African savannah. Considering the wide variety of sauropods, their extensive geographical distribution and their evolutionary success particularly throughout the Jurassic, it is difficult to see how one particular life-style could be applicable to all species. As Kurzanov has noted, there was probably a spectrum of habits exhibited by sauropods, ranging from terrestrial through to varying degrees of aquatic activity, especially for those frequenting floodplain areas. Many Mesozoic lakes and swamps were less than ten metres deep, and the most likely attitude for animals in such habitats was with the necks floating out on the water surface or inclined upwards, and with the tail dragging like an anchor. Fossilised trackways from lake beds show that swimming was sometimes assisted by kicking backward against the bottom with the clawed forelegs, much as hippopotami do, while a back foot touched the lake bed only when the animal made a turn. Tracks made on dry land prove that some sauropods were gregarious, herding animals, grouping in numbers of twenty or more with the adults on the flanks of the group protecting the very young and juveniles in the centre.

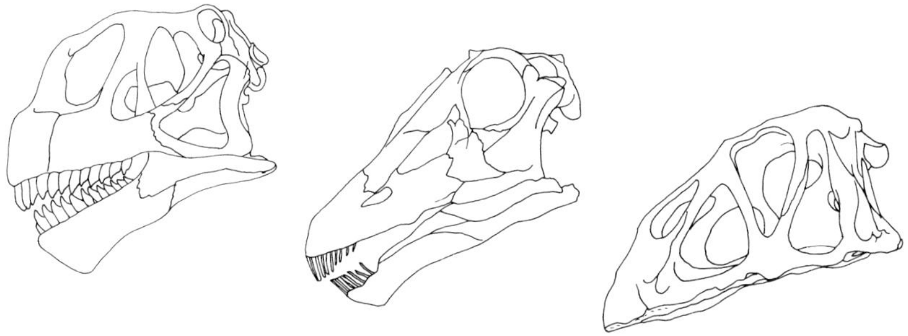

Part of the reason for the varying opinions on sauropod habits stems from the range of dentition evident within the group. The skull of Camarasaurus *(left), for example, is compact and has stout jaws supporting strong, deeply-rooted teeth throughout their length. The diplodocids such as* Diplodocus *(middle) and* Nemegtosaurus *in contrast, have longer, narrow skulls with feeble peg-like teeth only in the front section. Some sauropods such as the Chinese* Omeisaurus *(right) also had very fine teeth that are sometimes not preserved with the skull, though there are sockets for thirty-two in the upper jaw and between twenty-eight and thirty-four in the lower. (*Camarasaurus *based on Gilmore,* Diplodocus *based on Holland,* Omeisaurus *based on Zhiming)*

The most popular reconstructions of the largest of the dinosaurs are those of the Czechoslovakian artist, the late Zdenek Burian. Four paintings, one each of *Apatosaurus*, *Brachiosaurus* and *Diplodocus* done in the 1940s and 50s, as well as another of *Diplodocus* (dated 1965) show gregarious animals inhabiting shallow lakes and their shores. Apart from *Brachiosaurus* being depicted standing in water that we now realise is probably too deep, the reconstructions are very convincing. Burian's remarkable and prolific career spanned from the 1930s to the 1980s (working initially under the guidance of Prof. Josef Augusta and later under Prof. Zdenek Spinar), all of his numerous works having withstood the test of time unusually well, and requiring but minor alterations in the light of recent discoveries. For example *Apatosaurus* is shown with the blunt-nosed *Camarasaurus*-like skull rather than the more slender *Diplodocus*-like skull, the wrong skull having been used in the original reconstruction by O.C. Marsh (an error only corrected in 1975).

Some important new ornithomimosaur material, including three almost complete skeletons (two with skulls), was retrieved by the Polish-Mongolian Expeditions from the Nemegt Valley localities of Nemegt, Tsagan Khushu, and Altan Ula IV. In addition, a small skeleton was found by a Mongolian field expedition in 1967 at Bugeen Tsav outside the Nemegt Valley. These skeletons vary in size from the

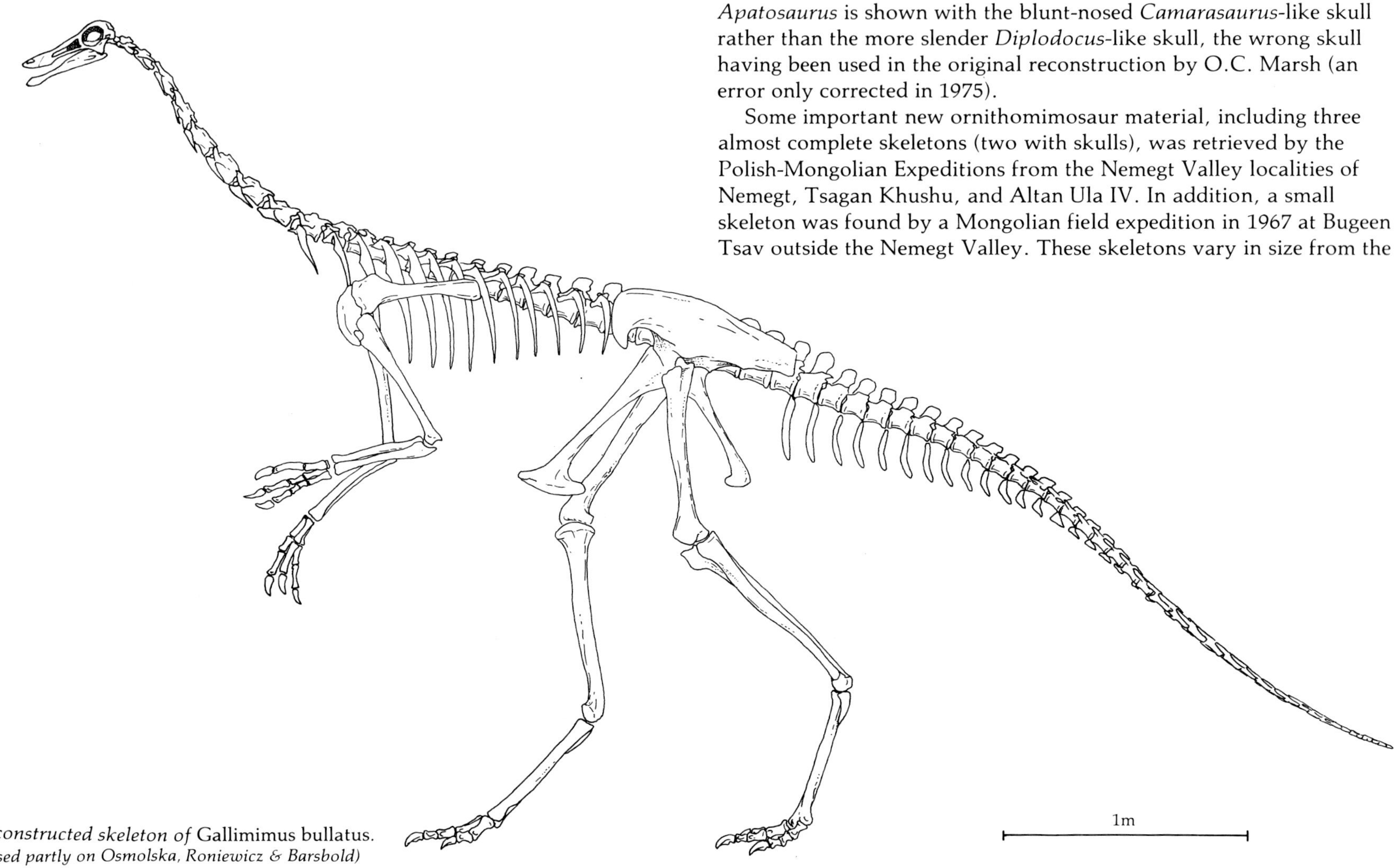

Reconstructed skeleton of Gallimimus bullatus.
(Based partly on Osmolska, Roniewicz & Barsbold)

shortest, with a skull length of 133 millimetres, to the longest with a length of 320 millimetres, and were the largest and best preserved ornithomimosaur remains found. All the material has been placed in the new genus and species *Gallimimus bullatus* (=chicken-mimic), adult specimens of which measured well over six metres in length and showed the typical body features that characterise this group of so-called ostrich-dinosaurs. The body was very shallow in breadth, being somewhat laterally compressed, and the lightly constructed tail and long, slender legs with three-toed, sharply-clawed feet all indicated an animal quite capable of outstanding cursorial abilities. These unusual animals appear to have been reptilian versions of large flightless birds such as the rhea and ostrich, and without doubt were at least as swift-footed. Groups of these lithe dinosaurs effortlessly running across the Mesozoic landscape in perfect coordination must surely have presented an incomparable sight.

Previously, it was believed that the hands of ornithomimosaurs were used for manipulation of prey prior to consumption, but the hands of

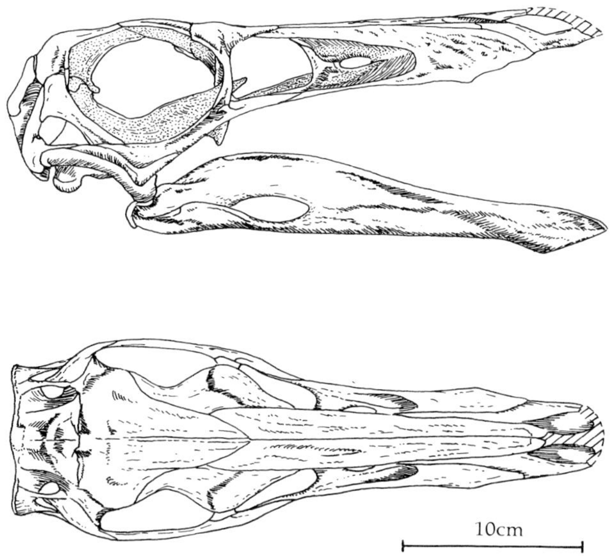

Reconstructed skull of Gallimimus bullatus. *Upper: Side view. Lower: Top view. (Based on G. Paul)*

Gallimimus appear to have been more suitable for raking or digging at ground level, possibly to expose potential insect or other invertebrate prey such as millipedes. Like *Saurornithoides*, the highly mobile neck and large, laterally directed eyes of *Gallimimus* were ideal for such activity. Thus Osmolska, Roniewicz and Barsbold envisaged this

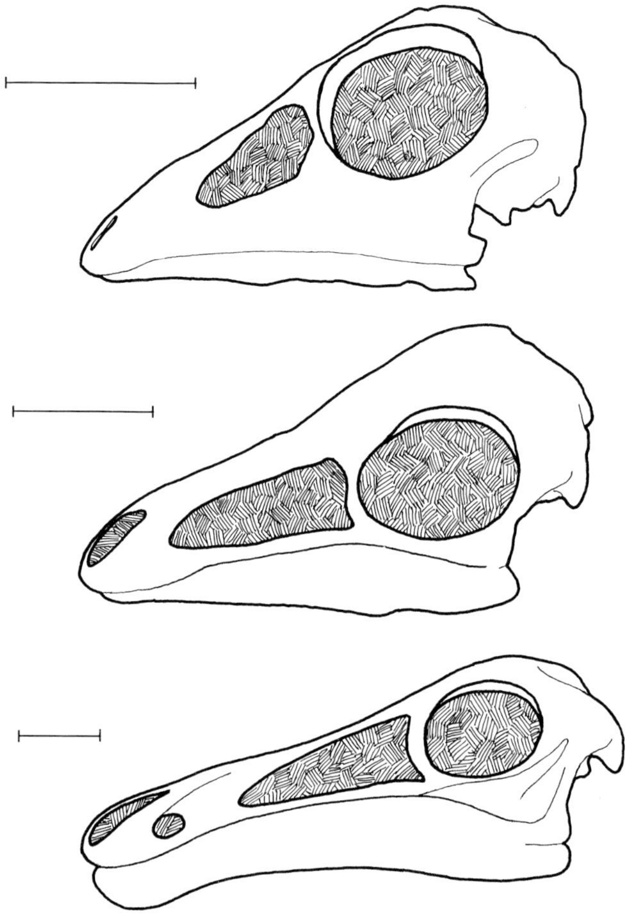

Progressive growth stages of Gallimimus bullatus *showing (top to bottom), a young juvenile, a half-grown specimen, and an adult. All scale bars represent five centimetres.*

Ornithomimosaurs

Ornithomimosaurs (=bird mimics), also known as the "ostrich dinosaurs" appear to have been the nearest dinosaur analogues to today's large flightless birds such as the ostrich and rhea. In approximate order of decreasing age, the Mongolian genera *Harpymimus, Archaeornithomimus, Garudimimus, Gallimimus,* and *Anserimimus* are known from early to late Cretaceous deposits, while the genera *Ornithomimus, Struthiomimus,* and *Dromiceiomimus* are known from mid– and late Cretaceous deposits in western North America. All shared a body morphology of a small bird-like head with large eyes and pointed snout, long, slender legs and arms with long fingers, and a light, balancing tail. Although classified within the Theropoda or "predatory dinosaurs", the skulls of ornithomimosaurs, which show little variation between genera, prove that these dinosaurs were not unlike giant herbivorous birds in appearance and habits.

With an adult length from six to 8.8 metres, *Gallimimus bullatus* (=chicken-mimic) is the largest, best preserved, and possibly the most evolutionarily advanced of the well known ornithomimosaurs. Remains of this species were discovered in the Nemegt Formation deposits, mostly within Nemegt Valley, during the Polish-Mongolian Expeditions between 1964 and 1967. The small, bird-like skull was slenderly constructed and permitted a relatively large degree of movement between adjacent bones. As in other ornithomimosaurs, the skull is light in construction yet is really more like the strongly-built skull of the extinct flightless New Zealand moa rather than that of the ostrich to which these dinosaurs are often compared. The jaws were toothless, elongated and spatula-like, being lengthened with sheaths of sharp-edged keratin similar to that of birds and turtles. The orbits were particularly large, and the brain's relatively small olfactory lobes and the low snout capacity in comparison to other theropods indicates that eyesight was the predominant sense. The brain cavity's size relative to the body was comparable to that of the ostrich, while the brain shape, middle ear and roof of the mouth were likewise very bird-like. As in *Saurornithoides,* the brain case of *Gallimimus* shows a cavity for an unusual pear-shaped bulbous structure, the function of which remains undetermined.

Gallimimus bullatus is one of very few theropod dinosaurs for which skeletons of different age groups are known, and as with the contemporary carnosaur *Tarbosaurus bataar,* some growth characteristics are noticeable. As *Gallimimus* aged, the relative size of the orbits decreased, the jaws lengthened, and the length of the posterior skull behind the orbits decreased, all trends that have similarly been observed in growing crocodiles. The forelimbs and tail became relatively longer with age, while the hindlimbs grew at the same rate as the body.

Since the discovery of fragmentary ornithomimosaur remains in North America in the 1890s, the probable life-style of these reptiles has received much attention but provided few conclusions. Originally, an insectivorous lifestyle not unlike that of the South American giant anteater was proposed by William Bebe. But traces of the necessary shoulder musculature and tongue structure were lacking, and after noting that most remains came from coastal areas, Barnum Brown proposed shoreline wading habits with an associated crustacean diet. Henry Fairfield Osborn rejected these concepts and provided the "ostrich theory" that described ornithomimosaurs as browsing herbivores capable of high speeds and using their forelimbs to bend branches to within reach of the beaks. William King Gregory agreed with Osborn but further suggested that the diet was omnivorous and may have included small vertebrates and large invertebrates.

The conclusions of Osborn and Gregory are now generally accepted, and ornithomimosaurs appear to have been fleet-footed omnivores capable of running at fifty kilometres per hour or more. They might have sought rolling steppes or open floodplains with scattered stands of low jungle cover where they could forage as well as use their speed and agility to evade predators. The exception to these concepts is *Deinocheirus mirificus* from Nemegt. Most authors accept that the ornithomimosaurs were the nearest relatives to *Deinocheirus,* though this does not give any clues to its life style. The muscles required to wield the powerful forelimbs of *Deinocheirus* recall Bebe's edentate theory to some extent, and it is possible that young deinocheirids were partly arboreal in habit, though the adults were of necessity fully ground-dwelling. Remains of various Mongolian ornithomimosaurs are housed at the Moscow Palaeontological Museum, the Polish Institute of Palaeobiology in Warsaw, and at the Ulan Bator Municipal Museum.

Opposite: Gallimimus bullatus, *based partly on G. Paul*

Left: *The slender, laterally compressed nature of* Gallimimus bullatus *is evident in an oblique view of the skeleton.* (*Photo by the author*)

dinosaur as feeding upon small, live prey swallowed whole, in a fashion similar to the modern galliforme birds such as the domestic chicken which exhibits a very similar neck bone structure. The extremely fast reflex capabilities of ornithomimosaurs would have enabled some prey such as large dragonflies to be caught "on the wing", while the diet probably also included a certain amount of soft plant material. Ornithomimosaur forelimbs were ideal for bending down small branches and twigs to bring fruits or young shoots and buds within reach of the bird-like beak, and large specimens could comfortably have fed on vegetation at a height of three metres.

The size range of *Gallimimus bullatus* skeletons collected allowed some conclusions to be drawn regarding ontogenetic changes that took place in this species. Post-embryonic growth was accompanied by a relatively decreasing orbit size (as one would expect), a progressive change in profile of the skull from somewhat triangular in very young specimens to slim and elongated in adults, relative increases in posterior skull size and snout size, and an appreciable increase in the length of the forelimbs compared to the hindlimbs (the radius increasing faster than the femur and tibia). The comparable proportions of the bones of the hindlimbs, however, changed only very slightly with age.

5.3. Dinosaur Enigmas

The most unexpected find of the 1965 expedition was made by Zofia Kielan-Jaworowska as she was exploring the Altan Ula III outcrops during a rainstorm on July 11. No doubt, walking through driving rain in one of the world's harshest deserts must have seemed suitably bizzare conditions for making an unusual discovery. From the track about a dozen large bones lay on a small hill of poorly-cemented sandstone outcrops, and scraping away the loose sand soon revealed an enormous thirty centimetre claw-like object shaped like a deep sickle. As night was approaching however, this obviously important find had to be left for the next day. That evening, the other expedition members back at base camp were incredulous at hearing the story of the giant claw, because at that time the enormous therizinosaur claws previously found by the Russian palaeontologists had not yet been recognised as dinosaur in origin, so there was no precedent for claws of such size.

The following day all the bones were removed. The full collection laid out upon the ground consisted of a shoulder girdle, two enormous tridactyl forelimbs and some isolated ribs. Unfortunately, all the other remains of this mysterious animal had disintegrated with erosion and time, and were irretrievably lost. Upon reconstruction, the giant forelimbs measured no less than 2.6 metres in length, and the only carnivorous dinosaur of sufficient dimensions to be compared with such remains were the tarbosaurs which grew to fourteen metres. But these giant predators had stunted, two-fingered forelimbs with claws never longer than fifty millimetres. Evidently, the new remains must have belonged to a very different type of carnivorous dinosaur because the rib size indicated a body similar to, or just slightly larger than, an adult tarbosaur, though with relatively much larger forelimbs.

Osmolska and Roniewicz designated this new find as *Deinocheirus mirificus* (=terrible hand), and erected a new family, the Deinocheiridae, lying within the superfamily Megalosauroidea (infraorder Carnosauria) to accommodate it. In morphology, the hands and forelimbs of *Deinocheirus* most closely resemble those of an ornithomimosaur, but it was not known if this was due to convergent evolution or to close evolutionary relationship. Osmolska and Roniewicz suggested that the deinocheirids may represent a link between the small and lithe coelurosaurs and the much larger carnosaurs though as outlined on page 94, recent study and comparison with other material now suggests that the deinocheirids appear more closely related to the ornithomimosaurs than to any other group.

After the completion of the 1965 season, field work by Polish palaeontologists was suspended for a year, but from 1967 to 1969, three small expeditions were organised to Bayn Dzak to continue the search for late Cretaceous mammals in the Djadokhta Formation strata. Although no major excavations were carried out during these seasons, a valuable collection of some twenty mammalian specimens was the highly-prized result of this work.

In 1970 and 1971, the second phase of the large-scale Polish-Mongolian Expeditions took place. As in 1963 to 1965, the main thrust of operations was again to be the exploitation of the dinosaur-bearing strata at Bayn Dzak and the Nemegt Basin and surrounding areas. Palaeontologists on the Polish side included Kielan-Jaworowska (1970, 1971), Teresa Maryanska (1970, 1971), Aleksander Nowinski (1970), Halszka Osmolska (1970), Hubert Szaniawski (1970), Adam Urbanek (1970), Andrzej Balinski (1971), Cyprian Kulicki (1971), and Andrzej Sulimski (1971). The Mongolian representatives included palaeontologists Rinchen Barsbold (1970), Demberlyin Dashzeveg (1970), Altangerel Perle (1970, 1971), and technical assistant Mr. Sambu (1971). As well, there was the usual complement of cooks, geologists, drivers, hired diggers, and a doctor.

Hypothetical life appearance of Deinocheirus.

The 1970 season included a two month period at the Nemegt Valley (late July to late September) during which were found two further incomplete skeletons of tarbosaurs, a complete skeleton of the hadrosaur *Saurolophus angustirostris*, an incomplete postcranial skeleton of the first lambeosaurine hadrosaur from the Nemegt Formation (named *Barsboldia sicinskii*), a postcranial skeleton and fragmentary remains of ornithomimosaurs, and fragments of coelurosaurs and numerous tortoises. However, it was finds of the unusual dome-headed pachycephalosaur dinosaurs, previously

Opposite page: *Replicas of the forelimbs of* Deinocheirus mirificus *from Nemegt displayed at the Warsaw Museum of Evolution. The claws are similar but not identical to the even larger ones from* Therizinosaurus (above, *also from Nemegt), whose size is evident when compared to a* Tarbosaurus *skull and forelimbs exhibited in the same display at the Moscow Palaeontological Museum. It is interesting to compare these remains with the giant ground sloth* Megatherium (right) *which may provide clues to the habits of such dinosaurs. Notice the large claws of the forelimbs. (Above courtesy of S.M. Kurzanov, Russian Palaeontological Institute, opposite courtesy of W. Skarzynski, Polish Palaeobiological Institute, right: exhibit at the British Museum of Natural History, photo by the author)*

unknown from this area and rarely found elsewhere, that was the most important material collected during these seasons. The Pachycephalosaurs are a very poorly known and perplexing group of ornithischians that had only been represented by fragmentary remains prior to the Nemegt finds. Historically, three genera were recognised, these being *Stegoceras* from North America, described by Lambe in 1902, *Pachycephalosaurus*, also from North America, described by Brown and Schlaikjer in 1943, and *Yaverlandia*, from the Upper Cretaceous of England, described by Galton in 1971. In addition, Bohlin described

DEINOCHEIRA
KOŚCI DEINOCHEIRA, TAK JAK
ZOSTAŁY ZNALEZIONE W AŁTAN UŁA
NA PUSTYNI GOBI, PODCZAS
POLSKO-MONGOLSKIEJ WYPRAWY
PALEONTOLOGICZNEJ W 1965 R.

very scanty material, probably belonging to *Stegoceras*, from Asian deposits in 1953. Because of the incomplete nature of the remains, opinions on the systematic position of the group ranged wildly, with different authorities considering them lying variously within the Stegosauria, the Ceratopia, the Ankylosauria, or the Ornithopoda. To add to the confusion, pachycephalosaurs are unique among dinosaurs in the possession of a pubis bone that is almost excluded from the acetabulum (hip socket), a condition shared with only one group of surviving archosaurs, the Crocodilia.

On account of the well preserved material of the new Mongolian genera, Maryanska and Osmolska were able to provide much needed information on the pachycephalosaur skull and post-cranial skeleton, as well as shedding light on the likely evolutionary position of the group. The Mongolian material consisted of four skulls, three with associated fragmentary skeletons, from the Barun Goyot and Nemegt formations at the Nemegt Valley locations of Khulsan and Nemegt. Four new genera were created to accommodate the finds which were designated as *Tylocephale gilmorei, Prenocephale prenes, Homalocephale calathocercos,* and *Goyocephale*. In contrast to previous skull finds from North America, which all appeared to show signs of wear as if having been rolled great distances by water, none of the Mongolian specimens showed any signs of transport. Thus they probably represent animals that were preserved in the areas where they died.

After considering the new specimens and comparing them to the North American material, Maryanska and Osmolska elevated the pachycephalosaurs to the rank of the new suborder Pachycephalosauria, lying within the Ornithischia and containing the single family Pachycephalosauridae. In addition to the previously mentioned genera, the genus *Stenopelix* (described by Meyer in 1859, and classified variously as either a hypsilophodontid or a psittacosaur), is also included in the Pachycephalosauridae because it shared their unique pelvic structure. As well as the characteristic thick-boned skulls, which varied in bony adornments and in height (from virtually flat to high-domed), other features of the pachycephalosaur skull included large, laterally-facing orbits tending to anterior vision, traces of large optic nerves indicating a high degree of visual perception, and extensive nasal cavities that provided keen olfactory abilities. Pachycephalosaurs were animals of primarily herbivorous habits, though a wide range of tooth abrasion patterns evident among the Asian genera imply differing varieties of shrub-level plant material in their diets. Considering the morphology of the legs, and the fact that the body was apparently unprotected by dermal armour, the authors suggest that these dinosaurs were comparatively slow moving and therefore may have used their domed heads as battering rams against would-be predators such as the many species of small Mongolian theropods common at the time. This is in contrast to Galton who had previously studied the North American material and concluded that pachycephalosaurs were somewhat more fleet-footed, and that the battering-ram behaviour was linked to the concept of sexual selection where males of some species of horned ungulates conduct head-butting competitions for the right to mate with females. Maryanska and Osmolska consider that pachycephalosaurs may have arisen from the hypsilophodontids in the early Triassic, an origin they possibly also shared with the Ceratopia, the two groups later separating in the Jurassic.

As mentioned, the remains of several ankylosaurid genera such as *Tarchia* and *Pinacosaurus* were collected on the 1970 and 1971 expeditions. Also during 1971, a particularly well-preserved new genus was found in the Upper Cretaceous beds at the Khulsan locality in the Nemegt Valley. The material consisted of a skull, the front portion of the trunk, shoulder girdle, forelimbs and sections of the armoury from an ankylosaur of considerable size. Named *Saichania* (saichan meaning "beautiful-one" in Mongolian), this animal grew to some seven metres in length, and is particularly notable for the inherent beauty of its patterns of armoury upon the dorsal carapace, skull, and about the neck (reconstruction on page 75). Since this original find other well preserved specimens of *Saichania* have been found, and it is now considered one of the best known ankylosaurs. The skull, which is virtually flattened on the dorsal surface, was strongly but rather elegantly protected with rounded bony scutes that fitted together in a cobblestone-like manner, while a multitude of conical spikes protruded from all round the neck region like a studded dog-collar. The back carapace and tail armoury consisted of symmetrical rows of spikes, the bases of which were surrounded by smaller, rounded scutes, these in turn being set within the bony matrix of the smoothly sloping back. As in other ankylosaurids, the long tail ended in a heavy, knobbed club that was probably used defensively should the attentions of a predator such as a tarbosaur become too persistent. *Saichania* had the most massive limb bones and the shortest metacarpals of any Asian

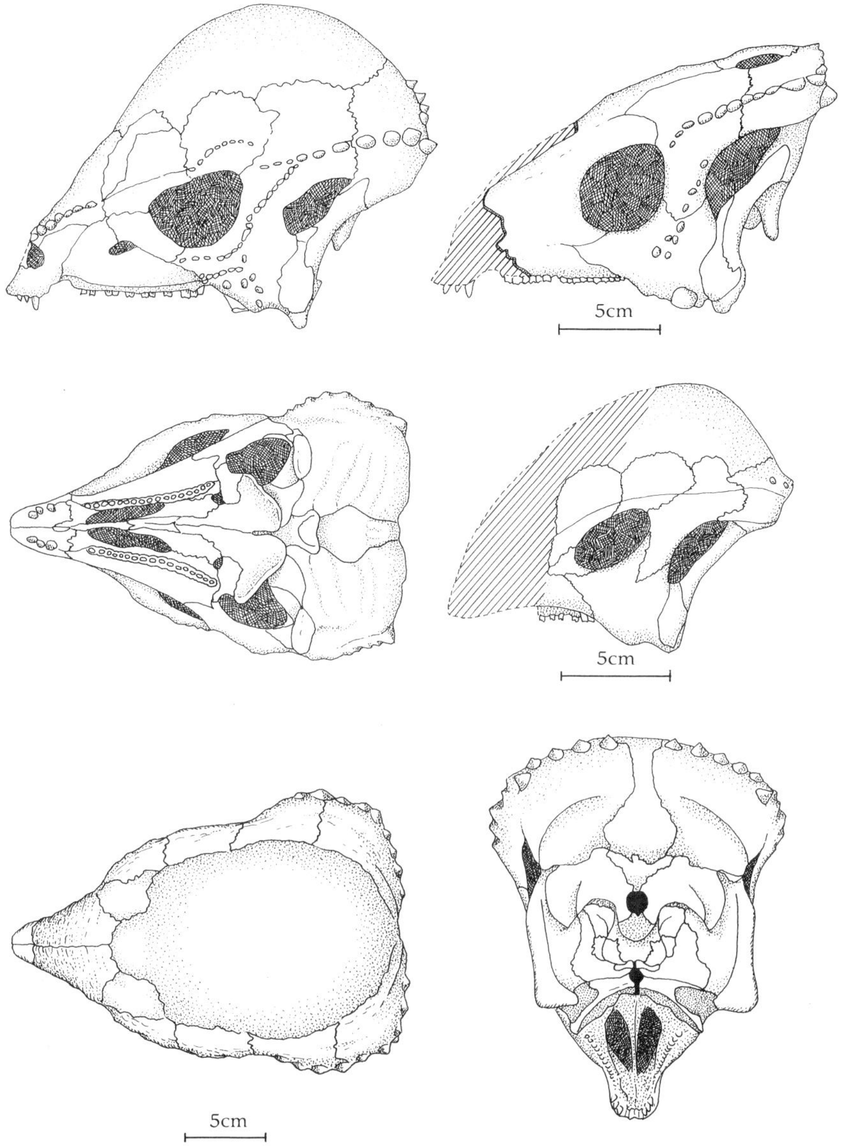

Anticlockwise from top left: *Skull of* Prenocephale prenes *in side, palatal, top, and rear views, and side views of the skulls of* Tylocephale gilmorei *and* Homalocephale calathocercos. *(Based on Maryanska & Osmolska)*

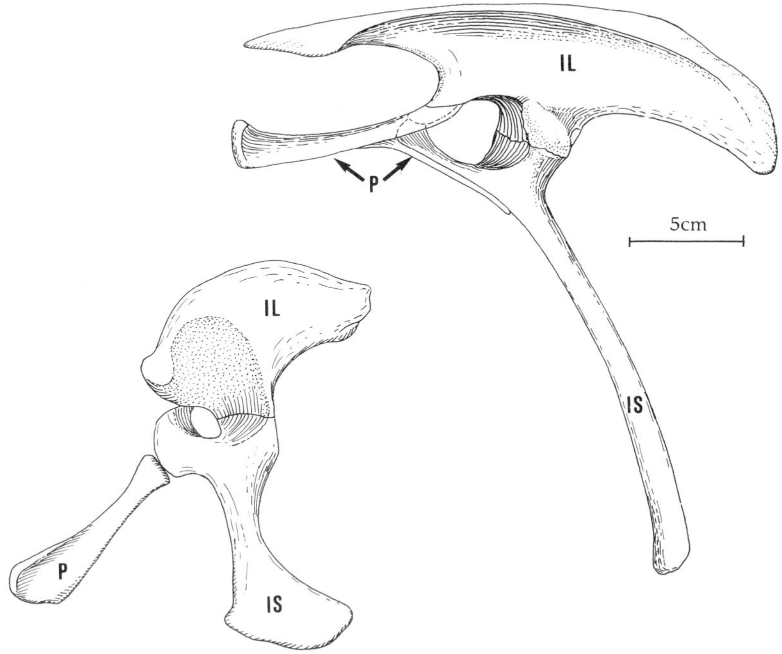

The pachycephalosaur pelvis is similar in some respects to that of crocodilians. In Homalocephale calathocercos (right), *the pubic bone hardly contacts the hip socket while in the alligator* (left, *not to scale*), *it is totally excluded.* (Homalocephale *based on Maryanska & Osmolska, alligator based on Romer) Key as for comparative diagrams on p. 21.*

ankylosaur, and thus achieved a very formidably armoured (yet not cumbersome) body propelled by powerful limbs.

Field work during 1970 and 1971 yielded a significant amount of protoceratopian material apart from the not unexpected collections of *Protoceratops andrewsi*. *Microceratops gobiensis* was found at the Sheeregeen Gashoon locality while *Bagaceratops rozhdestvenskyi* came from Khermeen Tsav, some forty kilometres southwest of the Nemegt Valley. *Microceratops* (=tiny horn-face) was originally described in 1953 by Bohlin, from remains in China. This was a small and unusually lithe animal, and unlike other protoceratopians that were quite slow

Pachycephalosaurs

The Pachycephalosauria, or "thick-headed reptiles", is perhaps the least well-known of the major dinosaur groups. Represented mostly from late Cretaceous deposits, about a quarter of currently recognised genera are Mongolian (including *Goyocephale, Homalocephale, Prenocephale,* and *Tylocephale*) while the others are from China (at least four genera) North America (about five genera), England (one genus) and Madagascar (one genus). Most of the Mongolian material was discovered during the 1970 joint Polish-Mongolian Expedition and is from the Barun Goyot and Nemegt Formations. It is particularly valuable not only for the thick, well-preserved skulls from which the group derives its name, but also for the preservation of post-cranial remains. This is a rarity for these animals, with body reconstructions having previously been largely speculative.

Prenocephale prenes (=sloping head) is known from a high-domed, bulging skull adorned with rows of small bony nodules along its sides and rear. *Tylocephale gilmorei* (=swollen head) had a similarly-shaped skull, though the snout and part of the dome is missing. *Homalocephale calathocercos* (=even head) had a flat, domeless skull that, as in *Prenocephale,* was adorned with rows of nodules along the rear and sides. Finally, *Goyocephale* (=decorated head) also possessed a relatively flat skull that had an irregular, pitted surface, and was armed with unusual pairs of long stabbing teeth in both jaws. All were quite small animals about the size of large dogs, and remains of the limbs, pelvis, tail, vertebral column and ribcage are variously preserved in *Prenocephale, Homalocephale,* and *Goyocephale.*

It is difficult to ascertain the systematic position of the pachycephalosaurs. The horn-covered pre-dentary bone of the lower jaw marks them as ornithischians, but their skull structure is unlike that of any dinosaur group. Some pelvic features appear remotely similar to the ankylosaurs (pachycephalosaurs having previously been classified within the Ankylosauria) but while the armour of ankylosaur skulls consists of new bone growing over existing skull bones, it is the original cranial bones that form the thick domes of pachycephalosaurs. Maryanska and Osmolska, who described the new Mongolian material, consider that pachycephalosaurs form a distinct ornithischian group and should constitute a suborder on the same rank as the ornithopods or ankylosaurs. They also suggest that the degree of dome height may be of useful diagnostic value in their classification.

Features of the skulls and post-cranial skeletons allow some comments on pachycephalosaur life-habits. Very large optic nerves and wide and laterally-facing orbits indicate keen vision, especially in *Prenocephale* where the angle of view extended forward to some degree, while the bony orbit walls offered excellent eye protection. The nasal cavities were generous and probably lined with mucus membranes to enhance their function. Although most pachycephalosaurs had similar dentition, differing wear patterns in the teeth of the Mongolian species prove that diets and therefore habitats were similarly varied. Food probably included leaves, twigs, fruits and seeds, as well as large, slow-moving insects and other invertebrates. The rounded margin of the premaxilla suggests that instead of an ornithischian-type 'beak', the tip of the snout was provided with an extensive soft lip which might have been utilised when feding on herbage. From the tibia/femur ratio of the hindlimbs, it is evident that these were not highly-cursorial animals. The hind-quarters were very broad (as was the body) suggesting that as is the case with boa snakes, pachycephalosaurs could well have been born live rather than hatched. The neck was short, the head most likely carried horizontally, and the tail was well-built, muscular, and largely inflexible.

Pachycephalosaurs might have been reptilian versions of modern ungulates such as bighorn sheep, living and feeding in social herds on well-drained upland areas. Galton supposed that the domed heads were used as battering rams by rival males asserting their dominance, in the same manner as many horned ungulates still do. It is more probable that pachycephalosaurs engaged in pushing test-of-strength contests rather than direct battering. The pelvic structure of *Homalocephale* shows that the long, low ilium contacts between six and eight vertebrae with ridged articulation joints, an ideal arrangement for transmitting powerful horizontal forces to the legs without causing injury. Considering the small size of their unarmoured bodies, it is also likely that male pachycephalosaurs used their heads for defensive purposes. When the herd was threatened, the female and young could retreat within a protective circle of larger males which were more capable of successfully confronting the danger.

Opposite: Homalocephale calathocercos

moving and quadrupedal, the long, slender hindlimbs of *Microceratops* suggest the capability of agile bipedal activity. Maryanska and Osmolska believe it is probable that a closely related animal was the ancestor of the true ceratopians that evolved into the spectacular variety of large gregarious dinosaurs roaming the North American plains in the late Cretaceous. Although the existence of similar large ceratopians in Asia has several times been suggested, the only evidence for them is what appears to be a fragmentary ceratopian horn core in the collections of the Moscow Palaeontological Museum. The remains of *Bagaceratops* (=small horn-face), another small protoceratopian, consist of fragmentary postcranial bones and a collection of at least twenty-two skulls from all age groups, ranging from forty-seven millimetres to twenty-five centimetres in length. Changes in the skull that accompanied growth were similar but not identical to those observed for *Protoceratops andrewsi*. They include a slight lengthening of the snout, a decrease in the relative length of the orbit, an initial increase in the length of the frill followed by a relative shortening at a later stage, and an increase in the frill width.

The final significant find of the 1971 expedition involved two familiar dinosaurs, but it was the unusual circumstance of their preservation rather than the animals themselves that was of most interest in this case. The discovery was made in August at the Toogreeg locality, just thirty kilometres west of Bayn Dzak whose "flaming cliffs" had been the scene of such excitement and incredulity on a hot July day in 1922. Now, the Polish workers uncovered two small skeletons, one a *Protoceratops*, the other a predatory *Velociraptor*. Both small dinosaurs had apparently been entombed together while locked in a poignant death-embrace. With the two small reptiles facing each other, the fore - and hind claws of the small *Velociraptor* firmly gripped onto the head shield of the similar-sized but heavier *Protoceratops*, while the "beak" of the *Protoceratops* had probably ruptured the other's abdomen. Some climatic or other phenomenon must have overtaken the two antagonists in their death throes, perhaps resulting in burial during an ashfall from a nearby volcanic eruption or from storm-driven sand. This fossil is considered significant because it was the first known example of mortal combat between two types of dinosaurs. It seems possible that the *Velociraptor* may have been attempting to plunder a *Protoceratops* nest when one of the adults was still in the vicinity. This is still conjecture, though it does seem a likely explanation.

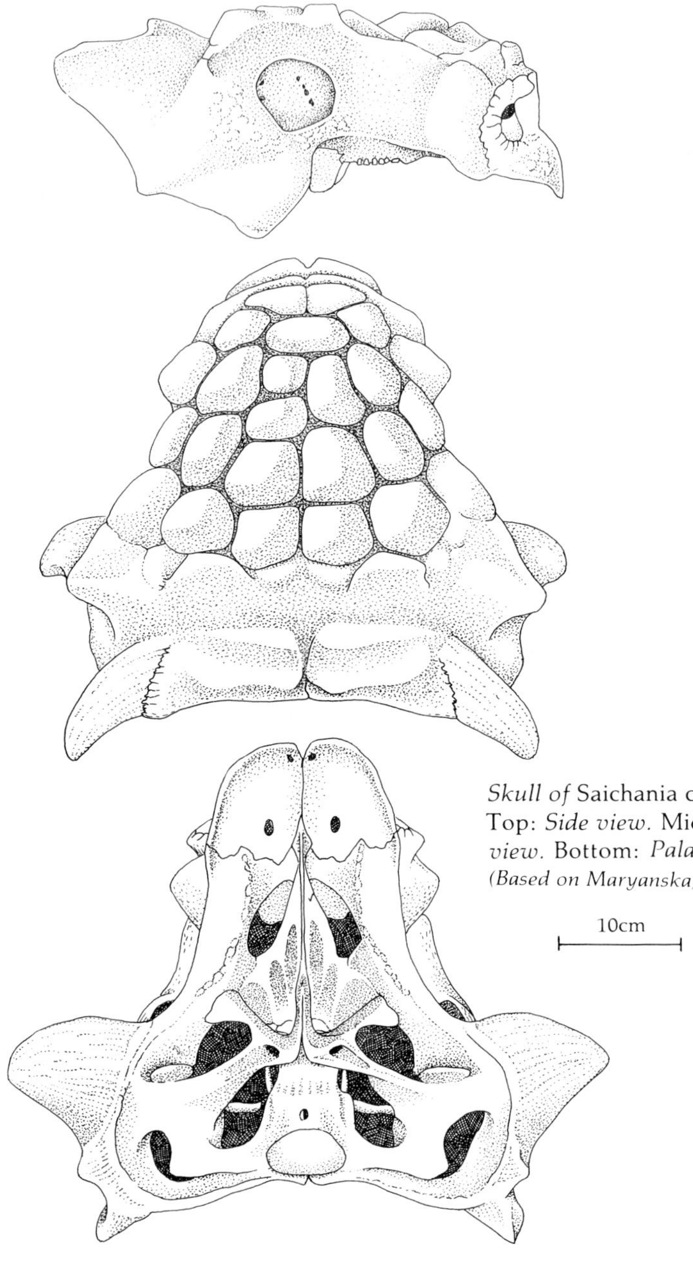

Skull of Saichania chulsanensis. Top: *Side view*. Middle: *Top view*. Bottom: *Palatal view*. *(Based on Maryanska)*

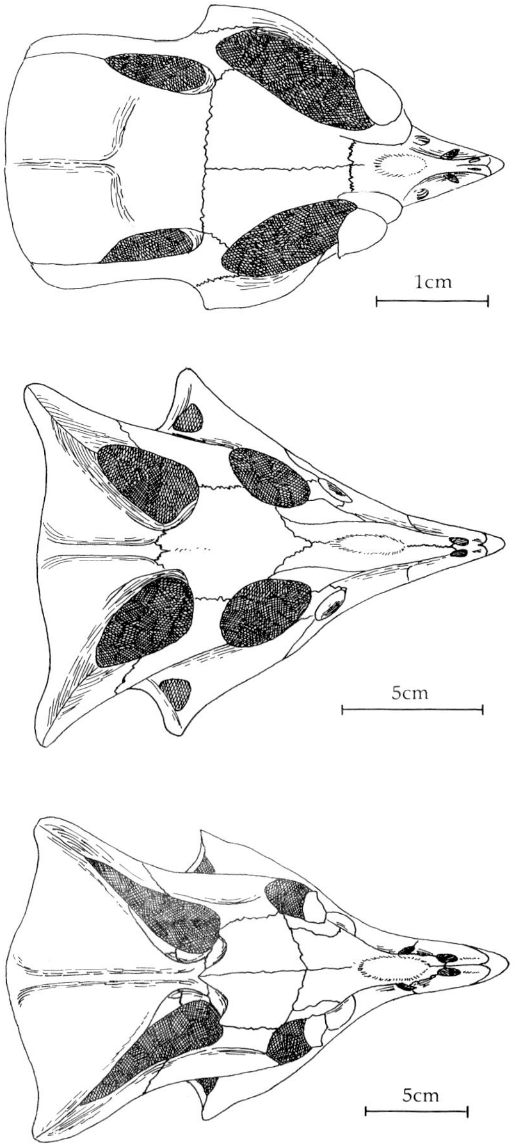

Progressive growth stages of Bagaceratops rozhdestvenskyi *showing (*top to bottom*), a hatchling, a juvenile, and an old adult. (Based on Maryanska & Osmolska)*

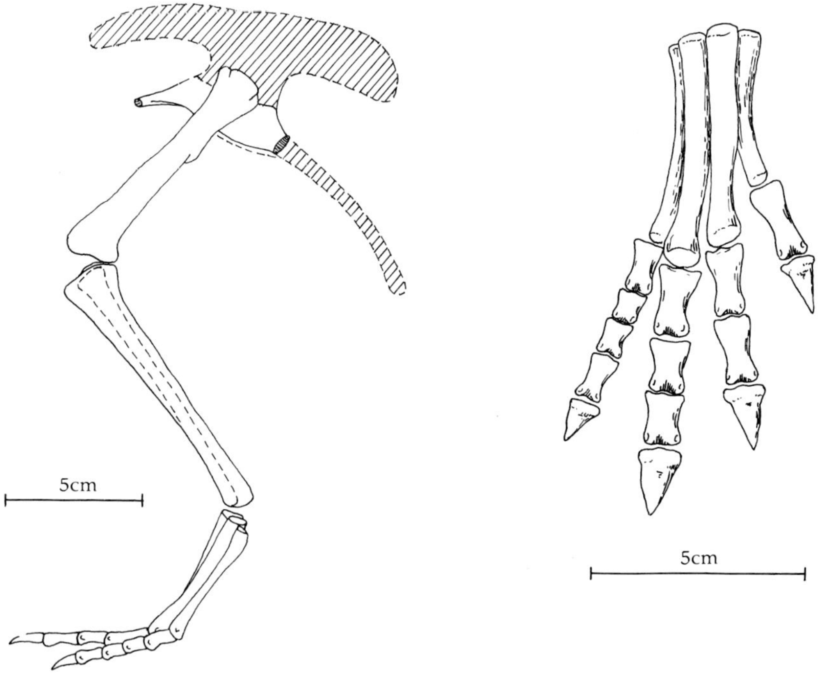

*Reconstructed pelvic girdle and hind leg (*left*) and foot of* Microceratops gobiensis. *(Based on Maryanska & Osmolska)*

These fossils were the last major finds of the Polish-Mongolian Expeditions, a fitting conclusion to a highly successful series. On August 12, the camp at Toogreeg was wound up, and the expedition members prepared for their return journey northwards to Ulan Bator which was reached a week later. Once the stone monoliths containing the fossils had been transported to the city, the entire collection was divided up for research purposes, half being retained at Ulan Bator for study by palaeontologists of the Mongolian Academy of Sciences, the other half being freighted back to Warsaw to the Polish Palaeobiological Institute. As with the Russian ventures of the 1940s, the Polish-Mongolian Expeditions had assembled a very comprehensive collection of dinosaurs, mostly from Nemegt but also from several other sites. In addition, significant collections of Cenozoic mammals had been made, particularly from sites in western Mongolia. But the

Above right: *Polish expedition members overlooking the Nemegt outcrops.*

Above left: *Excavation of a large tarbosaur skeleton at Nemegt during the 1970 expedition.*

Right: *Preparing the ankylosaur* Saichania *at Khulsan, Nemegt Valley, in 1971. (Courtesy of W. Skarzynski, Polish Palaeobiological Institute)*

most valuable non-dinosaur fossils were undoubtedly those of the tiny Cretaceous mammals of Bayn Dzak – shy, nocturnal animals that having survived for so long in the shadow of the larger reptiles, were eventually able to flourish and inherit their world.

Most of the technical results of the Polish-Mongolian field work are to be found in *Palaeontologica Polonica,* the official publication of the Polish Palaeobiological Institute. Yet it was not only laboratory researchers who were to come into contact with the dinosaur fauna of Mongolia as a result of this study. Dinosaurs were already somewhat familiar to Mongolians from earlier expeditions (and had even appeared on that country's postage-stamp issues), but in Poland, where non-marine Cretaceous fossils were non-existent, there was an extensive popular interest in the new discoveries. And so the Polish Institute of

Palaeobiology set aside an area of parkland at Chorzow in Silesia for the construction of a series of life-size dinosaur reconstructions. Designed by Wojciech Skarzynski and cast in natural attitudes among lakes, planted shrubbery and trees, they provide the public with imaginative and dynamic portrayals of many types of Gobian dinosaurs. Much of the skeletal material has also been placed on public display, with some of the complete skeletons having been assembled under the guidance of Zofia Kielan-Jaworowska to form part of the permanent display entitled "Evolution on Lands". This valuable collection, which includes many unique exhibits, is housed in Warsaw's stately Palace of Culture.

5.4. Desert Quest

Outside the three main periods of dinosaur collecting in the Gobi narrated in this and the preceding two chapters, there has been other similar work carried out from the late 1960s, sometimes even in cooperation with the Polish-Mongolian Expeditions. Primarily conducted by Russian and Mongolian workers, it has involved specialists such as Sergei M. Kurzanov, Konstantin E. Mikhailov, and T.A. Tumanova, A.F. Bannikov and A.V. Sochava on the Russian side, as well as Rinchen Barsbold and Altangerel Perle from the Mongolian side. Being of a less labour intensive and more protracted nature than the large-scale expeditions of earlier years, this work continues the long-term fossil exploitation programme of Mongolia. Many of the results can be found in *Paleontological Journal* and in the *Transactions of the Joint Soviet-Mongolian Palaeontological Expeditions* series. Although a full account of the field work is beyond the scope of this book, particular finds will be mentioned if they relate to topics discussed earlier.

Above: *Overlooking the Shiloost Ula site with the 1977 joint Russian-Mongolian Expedition camp toward the middle of the photograph. Ankylosaur remains and the nests of hadrosaurs were discovered here in Upper Cretaceous strata.*

Right: *Crating plaster-encased ankylosaur remains excavated from Upper Cretaceous strata at the Khongil locality in 1985.* Left to right *(standing beside the crate) are S.M. Kurzanov, D.O. Tschizhov and Y.M. Gubin. (Courtesy of S.M. Kurzanov, Russian Palaeontological Institute)*

The non-osteological material includes many nests containing eggs of ornithischian dinosaurs such as protoceratopians, hadrosaurs, and iguanodontids, as well as some from small theropods. Southern Mongolia is rich in late Cretaceous dinosaur eggshell remains, with the most important sites (apart from Bayn Dzak) being Khermeen Tsav, Tugrikiin Shireh, and Gurliin Tsav, while Builjasutuin Khuduk is one of the fewer early Cretaceous sites. Sochava compared various eggs collected from the northern Gobi with others from Kazakhstan and Provence in France to try and provide a morphological and possible phylogenetic comparison with the eggs of modern reptiles and birds. The eggs of living reptiles are divided into two groups: cleidoic, which are supplied from the beginning with all essential substances other than gases; and non-cleidoic, which although supplied with essential substances and salts, must take up the greater part of water from the outer environment. The eggs of modern crocodiles and birds can be classed as cleidoic, whilst those of turtles (testudinates) occupy an intermediate position between the Squamata (snakes and lizards) which produce non-cleidoic eggs, and crocodiles and birds. Under the polarising microscope, testudinate eggs show segregation patterns which impart strength to the shell and increase resistance to desiccation while also enabling the shell to expand to take in water. As outlined in Chapter 1.2, the complete liberation of the reptilian egg from this latter requirement permitted the development of the continuous calcareous shell characteristic of modern crocodiles, birds, and most dinosaurs, and is considered a major evolutionary achievement.

Sochava distinguished two general types of dinosaur eggshell described as either ornithoid, with a structure similar to that of birds' eggs, or testudoid which is similar to turtles' eggs. It is suggested that the two types may represent the two orders of dinosaurs, the ornithoid belonging to the Ornithischia, the testudoid to the Saurischia. Sochava noted that the eggs found in close proximity to the ornithischian *Protoceratops* are definitely of the ornithoid type, so much so that some French scientists have maintained that the eggs are avian in origin. But as mentioned previously, any possible relationship between dinosaurs and birds lies with the saurischians, not the ornithischians. If this is accepted, any similarities between ornithischian and avian shell must have been due to ecological constraints having had a greater degree of influence on shell development than did the lack of a close taxonomic relationship between the two groups. The alternative is that birds were derived at an earlier stage from the thecodonts (Chapter 1.4) and subsequently evolved in parallel with the saurischians, in which case they would not have shared particularly similar shell structures.

One find of historical interest was the first confirmed report of embryonic dinosaur bones within an egg, described by Sochava in 1972. The discovery was made while studying eggshells collected from late Cretaceous deposits of Senonian strata in the eastern Gobi. Although the egg concerned is ornithoid in structure, it differs from those previously described as ornithoid. Because the specimen had been subjected to recent weathering, most of the embryo was unfortunately lost, but the four small subparallel tubular bones identified correspond to the metatarsals I-IV of the foot, and appear similar to those of the protoceratopians *Leptoceratops* and *Protoceratops*. The degree of ossification suggests that the tiny embryo perished just prior to hatching.

Of the osteological material, the remains of late Cretaceous theropods are the most frequently encountered dinosaurs in the Gobi strata, and as noted by George Olshevsky, their variety is unmatched in any other fossil deposits. Two agile and competent predators that will be mentioned are *Alioramus*, a medium-sized tarbosaur-like carnosaur, and *Itemirus*, a small dromaeosaur-like carnivore. *Alioramus remotus* is from the Nogon-Tsav Formation and was described by Kurzanov in 1976 from a fragmentary seventy-centimetre-

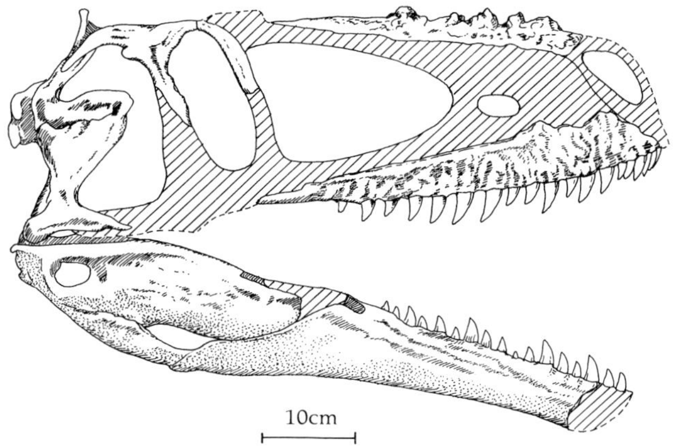

Skull of Alioramus remotus. *(Based on Kurzanov)*

long skull and incomplete skeleton estimated to be six metres in length. The relatively low snout is surmounted by a series of jagged irregularities even more pronounced than those of *Tarbosaurus*. In life, these probably supported a small nasal horn similar to but not as large as that of the North American genus *Ceratosaurus*. The lower jaw is very narrow and straight with a long dentary bone. These features, combined with the light structure of the skull's maxillary bone as well as other details, marks *Alioramus* as one of the least advanced of the tyrannosaurs. *Itemirus medullaris*, from Itemir in Dzhara-Kuduk, is known only from a short and high-domed braincase that is nevertheless quite distinct from those of other theropods. Described by Kurzanov in 1976, *Itemirus* was probably the size of *Velociraptor* or even smaller, and has very well developed optic lobes. Not unexpectedly, the inner ear indicates poor hearing but suggests an exceptional sense of balance, a prerequisite for the agile predatory activity undertaken by this dinosaur. *Itemirus* is considered to have been close to the shared ancestry of the carnosaurs and the smaller dromaeosaurs.

A number of new ornithomimosaurs have come to light during the Russian-Mongolian Expeditions, notable among them being the genera *Garudimimus*, *Anserimimus*, and *Harpymimus*. Described from late Cretaceous strata in 1981 by Barsbold, *Garudimimus brevipes* was the first of the ancestral ostrich dinosaurs to be found, and the remains include a virtually complete skull. Ancestral features are evident in the typical theropod-like cheek structure and the long, simply-constructed beak, while the foot still retains an unreduced central cannon bone as well as an inner toe. As in most ornithomimosaurs, the skull is toothless and exhibits the characteristic bulbous structure, though the end of the beak is notched and there is a small backward-projecting horn above and in front of the eye socket.

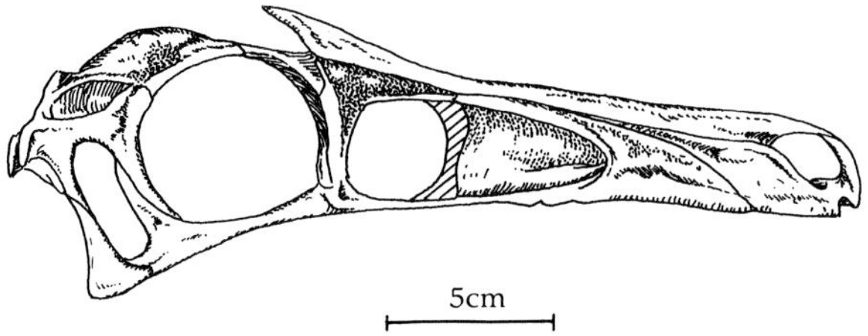

Skull of Garudimimus brevipes. *(Based on Barsbold)*

Unusual ornithomimosaur remains were found in 1981 in Shinehuduk Formation rocks (mid Cretaceous) in southeastern Mongolia, and described in 1984 by Barsbold and Perle as *Harpymimus okladnikovi*, the generic name being derived from the mythical Greek "harpyia" beast. This appears to be the most primitive of the ostrich dinosaurs and suggests a likely Central Asian origin for the group. *Harpymimus* is unique in that the tip of the lower jaw contains ten or eleven tiny conical teeth, rendering it a "toothed" toothless dinosaur. Although the ostrich dinosaurs were previously assumed to have evolved from dendate theropod ancestors, it was not until the remains of *Harpymimus* were found that there was any direct evidence to support this view. Other features of the skeleton are typically ornithomimosaur in structure, though the ratios of limb lengths and the pectoral and pelvic girdles differ from other genera.

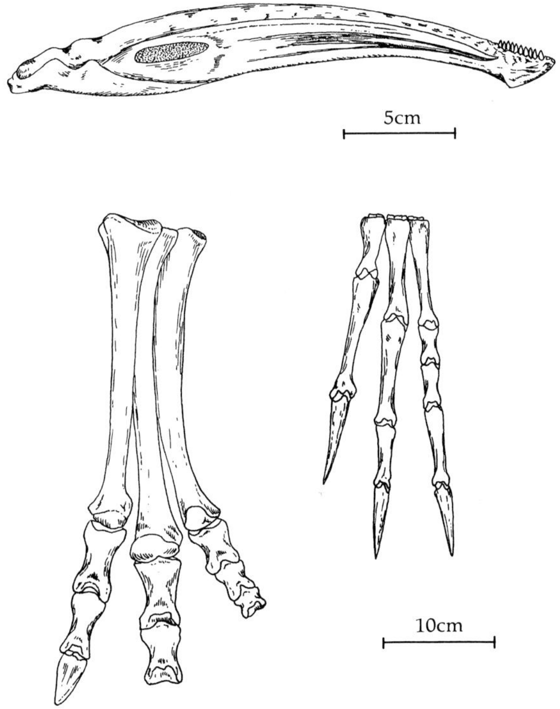

Top: *Lower jaw of* Harpymimus okladnikovi *(interior view)*. Bottom: *Foot (left) and hand of* H. okladnikovi. *(Based on Barsbold & Perle)*

Anserimimus planinychus is known from post-cranial remains collected from late Cretaceous rocks at the Bugin Tsav locality in southwestern Mongolia, and was described by Barsbold in 1988. The size of the scapula, shoulder joint, humerus and biceps muscle attachments all attest to the unusually powerful forelimb development in this animal, while the claws of the hand are unusually flattened, hence the specific name "flat-claw". Barsbold noted that although strong forelimb development was typical of the small predatory dinosaurs like the dromaeosaurs that used their arms offensively, *Anserimimus* was probably not such a predator. It was more likely to have used its claws to rake away loose substrate, rocks and riverside detritus and scrub, activities which would disturb and flush out small prey that could then be seized by the bird-like beak.

These and other remains show that the ornithomimosaurs evolved a far wider range of morphology in Central Asia than they did in North America, where only progressive types are known. Barsbold was able to list the Mongolian genera in a broad chronological series as 1) *Harpymimus* (early Cretaceous), lower jaw with some teeth retained and a so-called "prehensile" manus, 2) *Archaeornithomimus* (late Cretaceous), with a primitive non-compact manus structure, 3) *Garudimimus* (late Cretaceous), with retention of the first digit of the foot, 4) *Gallimimus* (late Cretaceous), a typical highly specialised representative very close to North American forms such as *Ornithomimus* and 5) *Anserimimus* (late Cretaceous), with relatively flattened claws and unusually strong forelimbs.

One group of particularly unusual Mongolian saurischians has direct bearing on basic questions of traditional dinosaur classification. So different are the segnosaurs from other dinosaurs that in 1980, Barsbold and Perle suggested that they be accommodated by the establishment of a new theropod infraorder, the Segnosauria (though in 1990 Barsbold and Maryanska suggested that segnosaurs may be more closely related to the Sauropodomorpha than to the Theropoda). Two genera and one indeterminate type of segnosaur have been described, these being *Segnosaurus* (=slow reptile) described by Perle in 1979, *Erlikosaurus* (=King of the Dead [Erlik] reptile) described by Perle in 1980, and a fragmentary specimen designated as the "dinosaur from Khara Khutul", which was generically unclassifiable. All three are late Cretaceous in age and are from deposits in southeastern Mongolia.

Segnosaurus galbinensis, from the Bayn Shireh Formation, was a medium to large quadrupedal animal with small skull, slender mandible and jaws that were toothless in the anterior section (most likely these were covered by a horny bill as in some ornithischians), while the teeth at the back of the lower jaw resembled those of the prosauropod *Plateosaurus*. Standing about two metres high at the pelvis, *Segnosaurus* had sturdy, tetradactyl clawed hindlimbs and tridactyl clawed forelimbs over half the length of the hindlimbs. *Erlikosaurus andrewsi*, from the Baysheen Tsav Formation, is known from a skull, some cervical vertebrae, a humerus, and both fore- and hind feet. The skull is similar to that of *Segnosaurus*, though it has a greater number of teeth and a larger beak, while the size of the claws indicate a smaller body size than *Segnosaurus*. The "dinosaur from Khara Khutul" is represented only by a pelvis which appears to be very similar to that of

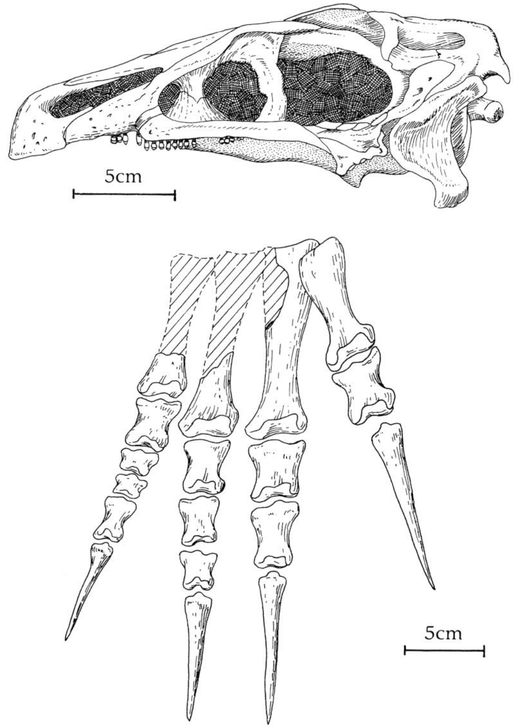

Reconstructed skull and right foot of Erlikosaurus andrewsi. *(Based on Barsbold & Perle)*

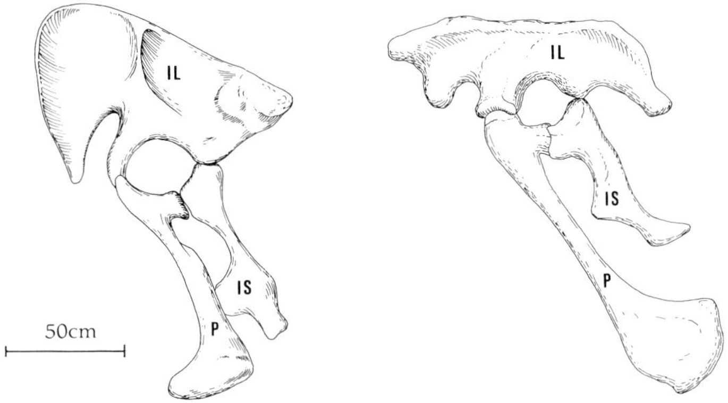

Pelvis of Segnosaurus galbinensis *compared to that of* Adasaurus *(right, not to scale). (Based on Barsbold & Perle) Key as for comparative diagram on p. 21.*

Segnosaurus though smaller in size. Because there is a possibility that this is a juvenile specimen of either *Segnosaurus* or *Erlikosaurus*, it has not been assigned a genus.

As noted in chapter one, it is the structure of the pelvis in the segnosaurs that is most unusual. Unlike other known saurischians in which the pubis is directed anteroventrally, the segnosaur pubis lies parallel to the ischium, just below the hip socket. This arrangement, which has been described as "opisthopubic" by Barsbold and Perle, seems closer to the pelvic structure of ornithischians than that of saurischians. Other parts of the osteology are also unusual, for example the possession of a horny beak covering the jaws which is an ornithischian trait, and the distinctly bird-like foot structure with the slender, closely positioned toes. The existence of the opisthopubic pelvis makes it somewhat more likely that *Archaeopteryx* may have had a posteroventrally directed pubis, rather than having it dislocated during preservation. More recently, a new type of Gobi dromaeosaur named *Adasaurus* has been found which also shows a segnosaur-like arrangement of the pelvic bones. Described by Barsbold in 1978, *Adasaurus* was late Cretaceous in age and appears to be very similar to *Velociraptor*. Like other Mongolian theropods such as the therizinosaurs and the deinocheirids, the segnosaurs appear to have deviated markedly from a typical saurischian morphology. Barsbold and Perle believe that the digits of segnosaurs were probably webbed and that these unusual dinosaurs might have had aquatic habits, feeding perhaps on fish caught by the sharp beak. Some authorities are sceptical that segnosaurs were agile enough to have fed in this manner.

Of the small theropod skeletal finds, two notable examples are *Avimimus portentosus* (= bird mimic) described by Kurzanov in 1981, and *Mononychus olecranus*, described by Perle et al. in 1993. *Avimimus* is known from a remarkable bird-like skeleton which shows a very slender, bipedal animal with long slim legs, bird-like feet, supple neck, and long tail. A number of avian characters of the skeleton suggest that it possessed plumage, the evidence being in the structure of the forelimb. The humerus shows a number of ridges similar to those of birds where they serve as attachment sites for the powerful flight muscles and wing folding mechanism. Unlike any other theropod, the ulna has a ridge along the rear edge which appears very similar to the series of bumps on the corresponding bone of the avian arm where they serve as the base of the secondary wing feathers. The skull is also very bird-like with advanced features such as a dome-like roof with traces of rich supplies of blood vessels (indicating good development of the cerebral hemispheres), lowered quadrate bone, backward movement of the nasal apertures, and the tendency toward an avian-like kinetic mechanism. Kurzanov envisaged *Avimimus* as having been a highly

Probable life appearance of Segnosaurus.

active bird-like late Cretaceous theropod that fed on flying insects. In pursuing prey such as dragonflies, it may have leaped into the air with the aid of its feathered arms, an activity which would have also helped it evade predators. As outlined in chapter 1.4, this is one of the theories advanced to explain the early stages of true flight.

Mononychus olecranus (= one claw) was discovered during the 1987 Russian-Mongolian Expedition, with a second specimen being found in 1992 during cooperative field work between Mongolian scientists and the American Museum of Natural History (the first excavations in Mongolia by the Museum since the 1920s). *Mononychus* was not capable of flight but may well have been a descendant of flying forms, and points to an unexpected diversity among primitive bird-like theropods. Being one of the best preserved of these, *Mononychus* had a keeled sternum and greatly reduced fibula which suggests it was more

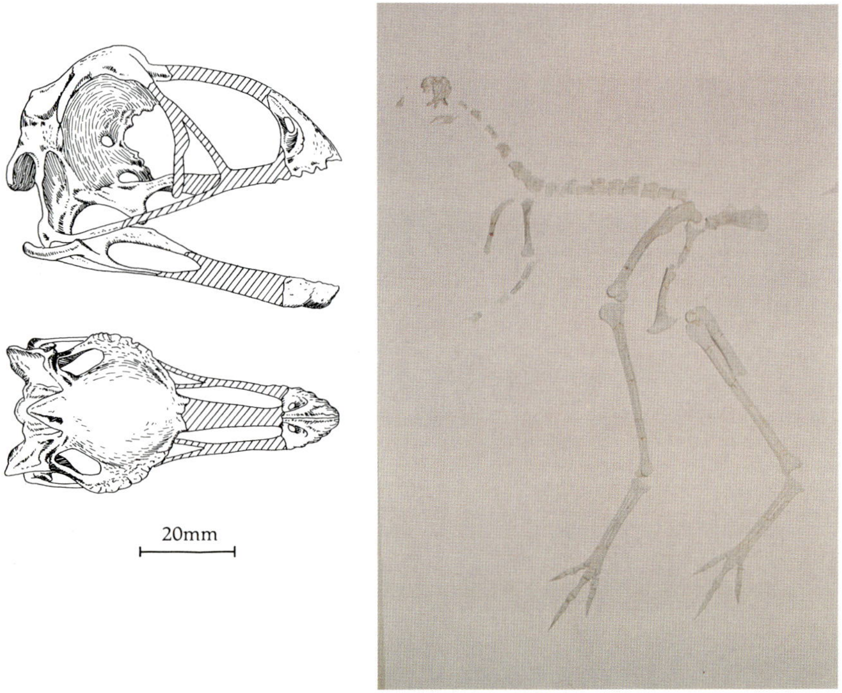

Reconstructed skeleton with side (upper) and top views of the skull of Avimimus portentosus. *(Based on Kurzanov)*

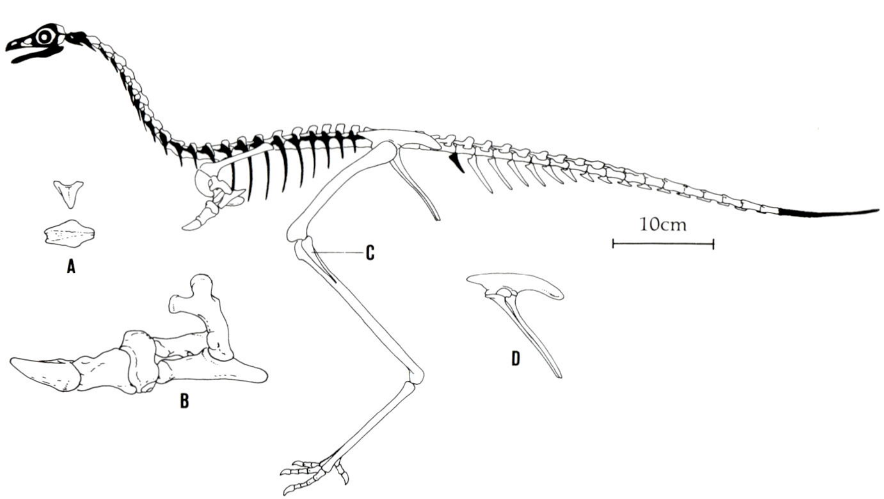

Composite reconstructed skeleton of Mononychus olecranus *with unpreserved elements shaded. A = sternum, in front (upper) and ventral views; B = detail of forelimb; C = reduced fibula; D = opisthopubic pelvis. Scale bar refers to the 1987 specimen, A B & D not to scale. (Based on Perle, Norell, Chiappe & Clark)*

closely related to modern birds than was *Archaeopteryx*. A small tooth preserved in one specimen indicates a fine dentition, while the pelvis is opisthopubic in design and finely built. Although most of the skeleton is typically avian, the stunted forelimb is similar to those of some burrowing animals, and ends in a single and unusual short claw on the first digit.

A further find relating to bird origins was described by Elzanowski and Wellnhoher in 1992, though it was actually found during the 1965 Polish-Mongolian Expedition to Bayn Dzak. The twenty-four-millimetre-long skull with widely-spaced teeth is from a juvenile theropod and presents a unique combination of avian and theropod characters. Named *Archaeornithoides deinosauriscus* (*Archaeornis* being a junior synonym of *Archaeopteryx*) this type is from the Djadokhta Formation (as is *Mononychus*) and is described as the nearest theropod relative to ancestral birds such as *Hesperornis* and *Archaeopteryx*. Some features of the skull are similar to those of troodontids, and it has been placed within a new theropod family, the Archaeornithoididae.

With respect to the Ornithischia, new Gobian ankylosaurs that have been found include *Amtosaurus* (= Amtgay reptile) from Bayn Shireh, which appears to be very similar to *Talarurus* (previous chapter), and

Shamosaurus scutatus (=Gobi reptile) from the Barun Bayn Formation of the Khamrin Us locality. Described in 1983, *Shamosaurus* is from early Cretaceous strata and had a low, broad skull with typical bony protuberances behind the cheeks. Together with *Sauroplites* (=hoplite reptile), described in 1953 by Bohlin from early Cretaceous rocks in Gansu, China, it is the oldest of the known ankylosaurs. Evidence of ankylosaur social behaviour has been provided by a unique 1988 find made during a four-year series of joint Chinese-Canadian Gobi expeditions. The remains of twelve ankylosaurs were found having possibly been buried together during a sandstorm, the similar sizes of the skulls of five of them (about twenty centimetres long) suggesting that although adults may not have been gregarious, young ankylosaurs of similar age grew up together and may have remained with the parents during this time.

Among iguanodontids uncovered by Russian workers are *Probactrosaurus mongoliensis* and *Iguanodon orientalis*. *Probactrosaurus* (=before *Bactrosaurus*) is known from incomplete skulls and reasonably well preserved skeletons that indicate both bipedal and quadrupedal activity. The osteology appears very similar to that of the Belgian *Iguanodon*, and it may even have been the same genus. The *Probactrosaurus* exhibit in Moscow is especially interesting because it includes nests of large, unhatched eggs believed to be from this species. *Iguanodon orientalis* is known from skull remains that also show some similarities to the European *Iguanodon*, though it is quite distinctive due to a greatly enlarged bulbous structure on the snout. Although most other iguanodontids lacked such a feature, the skull of the Australian genus *Muttaburrasaurus* exhibits a raised snout behind the nostrils, albeit nowhere near as large as that of the Mongolian species. The function of such structures is unknown, but they might have been linked to sexual selection in the same manner as the horns and antlers of modern ungulates.

Skull of Iguanodon orientalis. *(Courtesy of S.M. Kurzanov, Russian Palaeontological Institute)*

Finally, a 1987 joint Russian-Mongolian expedition to the Khermeen-Tsav locality in Nemegt Valley found three small protoceratopian skulls that Kurzanov later identified as identical to those tentatively described as *Protoceratops kozlowskii* by Maryanska and Osmolska in 1975. Examination of the new material proved that although closely related to *Protoceratops*, the remains differed sufficiently to be placed in a new genus designated as *Breviceratops* (=short horn-face). There is a range of skull sizes known for *Breviceratops*, varying from eight to thirty centimetres in length and from seven to thirty-two centimetres in postorbital width. Compared to *Protoceratops*, the new genus has a distinct horn, a lower and narrower neck collar, and a narrower preorbital section of the skull. As *Breviceratops* aged, the relative size of the orbits decreased (becoming less round and tending to

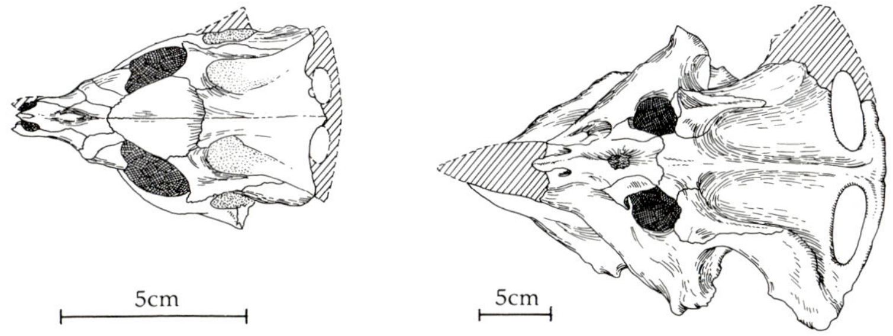

Skulls of a juvenile (left) and an adult Breviceratops kozlowskii. *(Based on Kurzanov)*

rectangular), the skull roof became flatter or concave and its length to width ratio decreased. The snout and bony collar both lengthened relative to the skull, while the collar also grew faster in width and its angle of inclination to the skull roof increased. *Breviceratops* is from the Barun Goyot Formation and being stratigraphically younger than *Protoceratops*, appears likely to be its direct descendant.

This concludes our narrative of palaeontological exploration that has been conducted in Central Asia. The fauna of this region represented the pinnacle of dinosaur evolution by some measures, being rarely equalled with respect to diversity and adaptation. Of the currently recognised Cretaceous dinosaur groups for example, all bar the hypsilophodontids and the stegosaurs (most genera of which had already succumbed in the Jurassic), are represented here. Moreover, a number of types such as the deinocheirids, the therizinosaurs, the psittacosaurs, the oviraptorosaurs and the segnosaurs were highly distinctive, some undoubtedly being unique to the region. With respect to ankylosaurs, protoceratopians and ornithomimosaurs, the fossil record indicates that Central Asia was the area of their main development and from where they later migrated to North America via the Bering Straits. In addition to the diversity of dinosaur genera contained in fossil collections, the fact that they include species represented by different age groups makes them exceptionally valuable to palaeontologists. Good ontogenetic material is now known for protoceratopians, (*Protoceratops*, *Bagaceratops* and *Breviceratops*), psittacosaurs (*Psittacosaurus*), carnosaurs (*Tarbosaurus*), and ornithomimosaurs (*Gallimimus*).

The geological field work that has been conducted over a wide range of fossil sites within Mongolia, especially within the Nemegt Valley, has provided some insights into the Cretaceous climatic and geographical conditions. The Nemegt Formation Beds are classified into Lower (early) and Upper (late) strata, showing a gradual geological transition between the two, though the Lower Beds tend to be more characteristically lacustrine (derived from lake environments) than the Upper which are more fluvial (river) in origin. Similar petrographies exist for both sets of beds however, with sands and clays as the predominant sedimentary material having apparently originated from areas some 100 kilometres to the northeast, according to Gradzinski. In the Cretaceous, the depositional area of Nemegt extended well beyond the boundaries of the present basin and formed a flat, very broad alluvial plain dissected by one or more giant rivers described as meandering or braided. Numerous gently curved channels varying in width from several metres to several kilometres formed a network fanning from the main rivers, their depth not exceeding ten metres. By the late Cretaceous, the Upper Nemegt Beds show evidence of some seasonality due to alternating rainy and drier periods resulting in varying river depths and widths. At times of lower rainfall, some channels noticeably shallowed and created series of plunge pools.

The petrographic and faunal composition of all Nemegt beds show that climatic conditions were generally humid and warm, with an abundant amount of rainfall. Fossilised remains of fruit suggest that *Nyssoidea* was a predominant genus of shrubby vegetation along river and lake shores, while the genus *Bothrocaryum* is also represented but was less common. Today, the fruit of *Nyssoidea* species is eaten and dispersed by birds and rodents, and it can be assumed that the plant may have played a similar role for small animals, birds and pterosaurs of the Cretaceous. Scattered calcified wood fragments are numerous in the western Nemegt Valley, and the Altan Ula locality of the Dragon's Tomb has preserved sections of tree trunks five to eight metres long. Dense and tall forests certainly covered sections of the Nemegt Valley on elevated ground between major river channels, with conifers constituting a major floral element. Under such equitable and constant environmental conditions, the Mesozoic reptiles flourished. And although it is possible that various dinosaur types were to some extent typical of particular geographical areas, the amount of habitat variation resulting from the Nemegt's fragmentary drainage system permitted many ecologically competitive herbivores to live in close association both with each other and with large carnivores. Relying on agility and the dense undergrowth in which to feed and evade predators, lesser dinosaurs such as many of the smaller theropods were able to coexist in the same habitats, while in ponds and pools, numerous crocodiles snapped at passing turtles and other small reptiles.

The largest denizens of this environment were of course the great sauropods, and remains of *Opisthocoelicaudia*, *Nemegtosaurus* and *Quaesitosaurus*, albeit incomplete, hint at a significant and distinctive representation in this part of the world. Although one cannot preclude browsing on cycadophyte and shrub foliage in some types (such as the camarasaurs which had relatively strong dentition), without doubt many sauropods fed within the extensive low-level fern and horsetail marshes that were a cosmopolitan feature of Mesozoic floodplains. Gregarious and passive herbivores, they moved with elephantine deliberation through the profuse dicksoniaceous fern growth, their mud-encrusted bodies resembling monolithic sandstone promontories arising above the horsetails, with their snake-like necks and diminutive heads almost continually lowered amid the delicate foliage. When temperatures rose, or if predators or parasites became too bothersome, herds of these dinosaurs would retreat to the refreshing coolness of the lakes and rivers, where bulky quadrupedal animals were deceptively transformed into floating behemoths.

The wetlands habitat provided abundant food for iguanodontids like *Probactrosaurus* and *Iguanodon*, the large ornithopod dinosaurs always having been associated with lowlands, swamps and estuarine habitats with their associated flora of horsetails, ferns, cycads, and conifers. But the most important marsh inhabitants were many types of hadrosaurs, of which the large saurolophs were the most abundant members at Nemegt. Aquatic adaptions, duck-like bills, strong grinding dentition, and preserved stomach contents indicate that secondarily-derived peaty or 'gyttja' material constituted a main part of hadrosaur diets. This, together with the frequency of their remains from coastal plain deposits, leads us to define the fern marsh habitat as of great importance to this diverse group of Cretaceous dinosaurs. Like hippopotami, hadrosaurs were probably responsible for the eutrophication and increased productivity of freshwater vegetation. Intelligent and highly perceptive, with skilled nesting and parental abilities, they were the most vocal and active members of the lowlands, their resonating bellows, social interaction, and aquatic habits rendering them conspicuous members of the dinosaur community.

Among the non-aquatic animals, it was the tarbosaurs that rank as most notable in the Nemegt fauna, and they were rarely to be found away from large rivers and lakes. Far from the brutish and tyrannical killers that carnosaurs are so often portrayed as, these gracile and agile predators spent much of their existence in relative inactivity alongside herds of large herbivorous dinosaurs. Although feeding was a group event, particularly when the carcass was from a large dinosaur, we can only speculate as to social interactions otherwise engaged in. If the nests of smaller theropods are any indication, tarbosaurs would have been careful nest builders and diligent parents, rivalling the patient skills shown by crocodiles and alligators. In addition to medium-size carnosaurs like *Alioramus*, for which only incomplete remains are known, a host of other saurischians including troodontids, dromaeosaurs, oviraptorosaurs, segnosaurs, coelurosaurs and ornithomimosaurs could be encountered in variety at Nemegt and its surrounding environs.

Rolling, open steppe country was the preferred domain of the ostrich-dinosaurs, loping swiftly as antelope across the open plains, and resting or feeding amid scattered stands of low jungle and scrub. Of all small theropods, the dromaeosaurs were the most voracious, and they may have hunted down small to medium-sized prey in groups not unlike wolf packs, using their large claws to dispatch their quarry. The dense undergrowth around pools and lakes would have been the ideal haunts of lithely-built insectivorous theropods the likes of *Avimimus*. Due to inherent fragility, they have unfortunately not fossilised well, but fragmentary remains indicate a number of such creatures inhabiting various parts of Central Asia both in the Cretaceous and in the Jurassic. The arboreal and cursorial activities of these or related thecodonts might resolve key questions surrounding the origins of birds and the development of avian flight, and further finds are eagerly awaited.

Of even greater consequence are new discoveries of the tiny Gobian Mesozoic mammals, which included the earliest known species. When the sun dropped below the horizon and night descended, they were part of an evasive nocturnal fauna that quietly replaced the active diurnal one. Small, shrew-like animls that aroused and tentatively emerged from their shelters and borrows to forage in the cool moonlight, they needed to be ever wary of the large-eyed and particularly agile troodontids that likewise sought food during the hours following dusk.

At some distances from large bodies of water dwelt theropods of a less familiar nature. The deinocheirids and therizinosaurs were creatures whose complete morphology we can as yet only guess at. If they were of a character outlined previously, their habitats, like those of modern and extinct edentates, were in areas denoted by scattered trees, low jungle, and mounds of social insect colonies that are today typical of the extensive savannahs of East Africa and northeastern South America (though the ground cover of the Mesozoic differed, grasses having not evolved until well into the Tertiary). Once again, only more remains will prove or disprove theories proposed to explain the large forelimbs and claws left by these creatures.

The "Valley of the Dinosaurs" Park at the Silesian Zoo in Chorzow, Poland. Above: *The ankylosaur* Saichania. *Above right:* Tarbosaurus *attacking the sauropod* Nemegtosaurus. Right: Gallimimus *(left) and* Tarbosaurus. Opposite page: *General view of the grounds with* Saurolophus *to the right (foreground) and* Nemegtosaurus *in the centre. (Courtesy of W. Skarzynski, Polish Palaeobiological Institute)*

In areas supporting cycadophyte and sclerophyll shrublands were to be found most of the ornithischian dinosaurs; the protoceratopians, psittacosaurs, pachycephalosaurs and ankylosaurs, though some ankylosaur species with weaker teeth may rather have fed in the fern marshes. The variety of protoceratopians ranged from heavy genera like *Protoceratops* to the slender and lithe *Microceratops*. *Protoceratops* appears to have dwelt and built its ubiquitous communal nests amid low sand dunes in terrain east of Nemegt that, although described as semi-arid, was nonetheless close to bodies of freshwater and vegetation subject to reasonable levels of rainfall. The laterally compressed turtle-like beaks of *Protoceratops* were ideal for feeding on the common cycadeoid bennettites, gymnosperms that supported crowns of palm-like fronds on strudy trunks up to three metres high.

As well as tearing the large fronds from their bases, they could also pluck out the .5 metre-long cauliferous strobili that grew between. Smaller protoceratopians and psittacosaurs, with their weaker neck musculature, may have subsisted mainly on fruits such as the large and abundant *Nyssoidea*.

Ankylosaurs and pachycephalosaurs were ornithischians of less specialised dietary requirements, and the obligatory low level vegetation in which the ankylosaurs foraged must have provided a wealth of invertebrate food to supplement their fibrous intake. Such zoological dreadnoughts, which have no comparable analogues in today's fauna, were probably not as common as many other ornithischians, but the extensive range of their foraging habits left but few areas devoid of their presence. Varied herbivorous diets were also applicable to the pachycephalosaurs, these small and incompletely-known dinosaurs having traditionally being viewed as upland dwellers. The Mongolian remains, however, are evidence of animals that either ventured down or regularly foraged in the deltaic lowlands at Nemegt, taking advantage of the rich feeding grounds the area provided.

But such faunal diversity could not last indefinitely, and by late Cretaceous times, the forces that had been raising mountain chains and draining swamps and lakes throughout the continental areas of the world now began to affect Central Asia as well. When this area lay near to sea level, it enjoyed the benefits of coastal conditions, but increasing continental plate boundary conflict and consequent drift activity resulted in a relentless elevation of the topography towards that of the current geographical situation, a semi-arid to arid grassy plateau rising to 2000 metres above sea level, hundreds of kilometres from the ocean. At the end of the Cretaceous, palaeobotanical evidence reveals that the flora of Mongolia as well as that of North America underwent significant changes, with deciduous forests of *Metasequoia* and *Trochodendroides* expanding over areas of high dinosaur diversity. The poorly-drained and extensive fern marshes and less prolific cycadophyte shrubland flora that was of such major importance to the ornithischians, could not endure and was eventually to be eliminated. The sclerophyll shrubland flora, although not wiped out completely, was nonetheless greatly reduced in extent, only to reappear in different composition during the late Paleocene.

Against substantial habitat and vegetation changes, and a climatic regime of an increasingly arid, continental, and seasonal nature, the Gobian dinosaur community began its inevitable decline. Even though the giant sauropods managed to considerably outlast their counterparts from other areas of the globe, they too, together with all their crested and armoured companions, finally passed into the realms of extinction. The sun had set for the last time on the Mesozoic world, time having claimed a fauna the like of which would never again be seen. In a paradox of fate, the very tectonic and climatic factors that brought about the demise of the Dragons from the Dunes would later result in a progressive desertification of the Gobi Plateau, the subsequent erosion of which has brought back so many of their well preserved remains.

The Last Dragon

In the stillness of the night
the last of a mighty dynasty
lies upon its broad flank
at the edge of a moonlit silvered lake,
gentle waters of a tomb to be.
Like a great engine,
the breath of life comes heavy and laboured,
the rising and falling of the giant lungs
once powerful and constant,
now shallow and irregular.
The end will not come quickly,
for life blood is still being pumped
by a heart not made to falter.
Yet death is on the cool breeze
for the will to inhale life has passed.
To an insect chorus from the twilight world,
one last breath is exhaled,
the great scute-patterned flank subsides
and is still.
Now a faint splash,
as concentric ripples from a passing turtle
lap gently against the open muzzle,
and the fire-red eye in its empty death gaze.

Index

Page numbers in bold type refer to species or genera cited in captions.

Adasaurus, osteology, **127**
Aetosaurs, 9
Alioramus, 131
A. remotus, osteology, **124**,125
Allactaga bullata, 76
Alligator, 84,117
Allosaurus, 12,35
Altan Teli, excavations, 96,97,100,101
Altan Ula, excavations, 78,85
Altan Ula IV, excavations, 101,108
American Museum of Natural History, 39,40,128
Amphibian egg, development, 6
Amptosaurus, 74,128
Anatosaurus, 15,33,86
Anaximander of Miletus, 27
Anchisaurus, 13
Andrews RC, 37,39,40,66
Andrewsarchus mongoliensis, **63**,**64**
Andrias scheuchzeri, 31
Ankylosauria, 15
Ankylosaurids, 15,74,129,133,134
Ankylosaurus, 15,74
Anserimimus, 110,125,126
A. planinychus, 126
Anthracotherium, 96
Apatosaurus, **13**,35,108
Archaeopteryx, 20-**24**,25,26,127,128
A. lithographica, 20,**21**
Archaeornithoides deinosauriscus, osteology, 128
Archaeornithoididae, 128
Archaeornithomimus, 110,126
Archidiscodon, **34**
Archosaur evolution, 7,8,9
Aristotle, 27
Augusta J, 108
Australopithecus africanus, 66
Avicenna, 27
Avimimus, 13
A. portentosus, osteology, 127,**128**
Bactrosaurus, 15,86
Badmajapoff T, 43
Bagaceratops, 16,52,54,55,130
B. rozhdestvenskyi, 54,117,120
growth stages, 120,**121**
osteology, **54**,121
Bakker R, 106
Balinski A, 113
Baluchitherium, 92
Bannikov AF, 123
Barsbold R, 11,60,100,109,113,123,125,126,127
Barsboldia, 86
B. sicinskii, 113
Bayn Dzak, excavations, 45,46,71,101,113,128
Bayn Shireh, excavations, 70,128
Bebe W, 110
Benaratherium, 93
Bering Straits, 18,52,74,130
Berkey CP, 42
Birds,
egg development, 20,124
evolution, 20-26,127,128
flight origins, 23-26,128
Bock W, 24
Bohlin B, 114,117,129
Borissyak AA, 40,93
Borsuk-Bialynicka M, 100,101
Bothrocaryum, 130
Brachiosaurus, 13,**36**,106
Brachyceratops, 16
Breviceratops, 52,130
B. kozlowskii, osteology, **54**,**129**,130
Brontosaurus, 13, see *Apatosaurus*
Brontotherium, **43**
Brown B, 110,114
Bugeen Tsav, 58
Builjasutuin Khuduk, 124
Burian Z, 108
Burnet T, 29
Buthus martensi, 2
Caenagnathids, 11
Caenagnathus, 60
Camarasaurs, 130
Camarasaurus, 35,**107**
Camelus bactrianus, 76
Camptosaurus, 14,35
Canis lupis, **64**
Capra ibex sibirica, 76
Captorhinomorphs, 7
Carnosaurs, 12
Central Asiatic Expeditions of the American Museum of Natural history, narration, 39-66
Ceratopia, 16
Ceratopians, 16
Ceratosaurus, **12**,35
Chasmosaurus, 16
Chilotherium, 97
Coelophysis, **11**
Coelurosaurs, 11
Colbert E, 38,106
Colgate SB, 42
Compsognathus, 11
Conan Doyle Sir A, 78
Coombs Jr W, 55
Cope ED, 35, 36
Corythosaurus, 15,86
Cotylosaurs, 7
Crocodilians, 7,20,22,50,81,84,124
nesting habits, 81,84
Cuvier Baron G, 31
Cynognathus, 8
Dart R, 66
Darwin C, 21,31
Dashzeveg D, 100,113
Davitashvili LSh, 19
De Ricqles A, 19
Deinocheirids, 131
Deinocheirus, **113**
D. mirificus, 92,94,110,113,**115**
Deinonychus, 11
Deltatheridium, 63
Deltatheroides, 63
Desmatosuchus, 9
Dicraeosaurus, 36,102,105
Dimetrodon, 7
Dinosaur eggs, 7,20,23,47,56,124
Dinosaur extinction, 17-20,134
Diplodocus, 13,35,102,106,**107**
Dodson P, 52
Doedicurus, **76**
Dovchin N, 100
Draco, 24
Dragon's Tomb, 130
excavations, 85
Dromaeosaurs, 11,131
Dromaeosaurus, 11,60
Dromiceiomimus, 110
Dryosaurus, 14
Dyoplosaurus giganteus, see *Tarchia*, 71
Eaglon J, 78
Echidna, 39

Edaphosaurus, **7**
Efremov IA, 37,68,91
Egg Mountain, Montana, 88
Elasmosaurus, **6**
Elmisaurus, 13
Elzanowski A, 128
Eohippus, **97**
Equus przewalskii, 3
Erlikosaurus, 13,126
E. andrewsi, osteology, **126**,127
Euoplocephalus, 74
Eurhinosaurus, **10**
Fabrosaurs, 14
Fabrosaurus, 14
Feduccia A, 23
Flaming Cliffs, see Bayn Dzak
Flerov KK, 68
Galeodes, 2
Gallimimus, 12,110,126,130,**132**
G. bullatus, 108,109-**111**,**112**
growth stages, **109**,110,112
osteology, **108**,**109**,110,112
Galton P, 22,114,118
Garudimimus, 110,125,126
G. brevipes, osteology, **125**
Gazella subgutturosa, 76
Genghis Khan, 2
Gobi Desert
geography, 1-4,18
geology, 130
Gondwanaland, 9
Goyocephale, 17,116,118
osteology, 118
Gradzinski R, 100,130
Granger W, 41
Gregory WK, 47,110
Gromova V, 93
Gurliin Tsav, 124
Haberlein E, 21
Haberlein K, 20
Hadrosaurs, 14,15,86
nesting habits, 88,89
Haloxylon, 78
Harpymimus, 110,125,126
H. okladnikovi, osteology, **125**
Hecht M, 22,23
Heilmann G, 22,23
Hennig E, 36
Herodotus of Halicarnassus, 27
Hesperornis, 128
Heterodontosaurs, 14
Heterodontosaurus, **14**
Hipparion, **97**
Hitchcock E, 33
Homalocephale, 17
H. calathocercos, 116,117,118,**119**
osteology, **117**,118
Homo diluvii testis, 31
Hooke R, 29
Horner JR, 88
Huxley TH, 21,22
Hylaeosaurus, 15
Hyotheridium, 63
Hypsilophodon, **14**,54
Hypsilophodontids, 14
Hyracotherium, **97**
Ichthyosaurs, 7,9,10,18
Iguanodon, **14**,36,129,131
I. orientalis, **129**
Iguanodontids, 14
Indricotheres, 92,93,96
Indricotherium, 93
I. transouralicum, 92,**93**
osteology, **92**
Inner Mongolia, 64,67
Irdin Manha, 63
Iren Dabasu Basin, excavations, 42
Itemir, 125
Itemirus medullaris, osteology, 125
Janensch W, 36
Jugoslavia, 96
Kannemeyeria, 8
Kazakhstan, 24,25,40,63,92,93,96
Kazmierczak J, 100
Kentrosaurus, 15,36
Kermack K, 106
Khamrin Us, 129
Khara Khutul, 126
Khermeen Tsav, 117,124,129
Khulsan, excavations, 116
Kielan-Jaworowska Z, 5,37,63,99,100,112,113,123
Kirgizia, 22,24
Kowalski K, 100
Kozlowski R, 99
Krausel R, 86
Kubiak H, 100
Kublai Khan, 3
Kulicki C, 113
Kurzanov SM, 52,107,123,124,125,127,129
Lamarck JB de, 31
Lambe LM, 114
Lambeosaurus, 86
Larsen FA, 42
Laurasia, 9,18
Lefeld J, 100
Leonardo da Vinci, 27,28
Leptoceratops, 16,52,124
Lesothosaurus, 14
Lhuyd E, 29
Lister M, 29
Longisquama, 9
L. insignis, 22,**23**
Lufengosaurus, 36
Lycaenops, **8**
Lystrosaurus, 8
Maiasaura, 88
Malecki J, 100
Maleev EA, 68,71,73
Maleyev EA, see Maleev
Mamenchisaurus, 13,37,102
Mammals, 62-64,92,93,96,97,100,101,131
evolution, 8,9
Mammuthus primigenius, 34,**35**
Manakin Colonel, 36
Manchuria, 67
Mandschurosaurus, 36
Marco Polo, 3
Marmota sibirica, 76
Marsh OC, 23,35
Martin LD, 22
Maryanska T, 71,74,100,113,116,118,120,126,129
Matthew WD, 39
Megalosaurus, 12
Megapodidae, 84
Megatherium, **94**,**114**
Mesozoic Era, 6,17-20
Metasequoia, 134
Meyer H, 20,116
Microceratops, 16,52,132
M. gobiensis, 117,120
osteology, **121**
Mikhailov KE, 123
Mongolia, 1,2,67
maps, 40,66,98
Monoclonius, 16,35
Mononychus olecranus, osteology, **128**
Montanaceratops, 52,55
Mook CC, 47
Morris FK, 42
Mosasaurs, 18
Muttaburrasaurus, 14,129
Nemegt, excavations, 78,85,91,101,108,113,116
Nemegt Valley,
geography, 76,78
geology, 91,130
Nemegtosaurus, 13,130,**132**,**133**

N. mongoliensis, osteology, 102,104,105,**107**
Nile crocodile,
nesting habits, 81,84
Nipponosaurus, 86
Nitraria, 78
Nodosaurids, 15
Nodosaurus, 15,35
Nopcsa Baron F, 23,72
Nothosaurs, 9
Novojilov N, 68,91
Nowinski A, 100,113
Nyssoidea, 130,133
Obruchev VA, 40
Olsen G, 47
Omeisaurus, 102,**107**
On Gong, 64
Ondai Sair, 64
Opisthocoelicaudia, 13,130
O. skarzynskii, 101-**103**
osteology, 102,104,**106**
Orlov AI, 67,68
Ornithischians, 7,14-17
Ornitholestes, 11,23,59
Ornithomimosaurs, 11,12,131
Ornithomimus, 12,35,110,126
Osborn HF, 39,66,106,110
Oshih, 64
Osmolska H, 60,100,109,113,116,118,120,129
Ostrom J, 20,22,23
Ouranosaurus, 14
Outer Mongolia, 67
Oviraptor, **11**,37,55,56,59,**60-62**
O. philoceratops, 60,**61**,62
osteology, 59,60,**62**
Oviraptorosaurs, 11
Ovis ammon, 76
Owen Sir R, 21
Pachycephalosaurs, 16,17,133,134
Pachycephalosaurus, **16**,114
Padian K, 24,25
Pangaea, 9,10
Paraceratherium, 93
Paranecellodidae, 85
Parasaurolophus, **15**,86
Paul G, 59
Pelycosaurs, 7,8
Pentaceratops, 16
Perle A, 22,113,123,125,126,127
Phytosaurs, 9
Pinacosaurus, **15**,73,74
P. grangeri, **71**,72
osteology, **72**
Placodonts, 9
Plateosaurus, **13**
Plato, 29
Platybelodon, 64
P. grangeri, **64**
Plesiosaurs, 7,18
Pliny, 27
Podopteryx mirabilis, 24
Polacanthus, 15
Polish Institute-
of Palaeobiology, 99,123
Polish Institute-
of Palaeozoology, 99
Polish-Mongolian Gobi-
Expeditions, narration, 99-123
Pope CH, 42
Praeornis sharovi, 25,**26**
Prenocephale prenes, 116,117,118
osteology, **117**,118
Probactrosaurus, 131
P. mongoliensis, **90**
osteology, 129
Procheneosaurus, 86
Prokennalestes, 63
Pronin V, 85
Prosauropods, 13
Proterosuchians, 9
Protiguanodon, 54
Protoceratopians, 7,16,52,132,133
Protoceratops, **16**,33,37,47,**48**,**49**,**52**,**54**,55,**56**,57,60,71,
120,124,129,130,132
P. andrewsi, **37**,47,**48**,**51**,52,**53**, 56,**57**,117,120
growth stages, 50,51,54
nesting habits, 47-50,56
osteology, 50,51,52,54
sexual dimorphism, 52
P. kozlowskii, see *Breviceratops*
Przhevalski Colonel N, 3
Pseudosuchians, 9
Psittacosaurs, 16
Psittacosaurus, **16**,37,54,**55**,85,130
P. mongoliensis,
growth stages, 55
osteology, **54**,**55**
Pteranodon, **10**
Pteromys, 24
Pterosaurs, 10,20
Python sebae, **81**
Quaesitosaurus, 13,130
Q. orientalis, osteology, 102,**107**
Quetzalcoatlus, 10
Rautian AS, 25,26
Reptilian egg, evolution, 6,7,124
Rhacophorus, 24
Rhamphorhynchus, 10
Ristoro d'Arezzo, 28
Roniewicz E, 109,113
Rozhdestvensky AK, 37,68,85,86,94,96
Russian Academy of Sciences'-
Gobi Expeditions,
narration, 67-97
Russian-Mongolian Gobi-
Expeditions, 123
Russian Palaeontological-
Institute, 67,68
Saichania, 15,74,**75**,116,117,**122**,**132**
osteology, 116,117
S. chulsanensis, osteology, **120**
Saltoposuchus, 9
Sambu Mr, 113
Saurischians, 7,11-13
Saurolophus, 15,37,**85**,**133**
S. angustirostris, 85,86,**87**,88,**89**,**90**,**91**,113
osteology, 85,86,**88**
S. osborni, 86
Sauroplites, 15,74,129
Sauropodomorpha, 13
Sauropods, 13,102,105-107,130,131
Saurornithoides, 37,56,**58**,59
osteology, 55,58,**59**
Saurornithoidids, 11
Scheuchzer JJ, 30,31
Schizotherium, 96
Schlaikjer EM, 114
Scolosaurus, 72
Scutellosaurus, 14
Seeley HG, 21
Segnosauria, 126
Segnosaurus, 13,126,**127**
S. galbinensis, osteology, 126,**127**
Shabarakh Usu, see Bayn Dzak
Shackelford JB, 42
Shamosaurus, 15,74
S. scutatus, 129
Shantungosaurus, 36
Sharov AG, 22,24,25
Sharovipteryx, 24,25
S. mirabilis, **25**
Shrendyb Prof, 99
Smilodon californicus, **34**
Sochava AV, 123,124
Sordes pilosus, 24
Spinar Z, 108
Spinosaurus, 12

Stegoceras, 17,114,116
Stegosaurs, 15
Stegosaurus, 15,**35**
Steno N, 29,30
Stenonychosaurus, 11
Stenopelix, 116
Stenson N, see Steno
Stigler R, 106
Struthiomimus, **12**,110
Styracosaurus, **16**,55
Sulimski A, 100,113
Syrmosaurus vimicaudus,
see *Pinacosaurus grangeri*, 71
Szaniawski H, 113
Talarurus, 15,**73**,74,**75**,128
T. disparoserratus, 71
T. plicatospineus, **70**,**71**
Tarbosaurus, 12,37,80,**82**,83,**84**,85,94,**114**,130,131,**132**
T. bataar, **17**,79,80,82,**83**,**90**,**96**,101
feeding behaviour, 81
growth stages, 81
nesting habits, 84
osteology, 79,**80-82**
Tarchia, 15,71,74,**75**
T. gigantea, osteology, **72**
Tarsitano S, 22,23
Taxartosaurus, 86
Tenontosaurus, 14
Tethys, 9
Thecodonts, 7,9,21
Therapsids, 8
Therizinosaurs, 95,131
Therizinosaurus, 13,95,**114**
T. cheloniformis, **92**,94
Thylacosmilus, **76**
Tochisaurus, 11
Toogreeg, excavations, 120
Torosaurus, 16,55
Trachodon, 35
Transbaykalia, 92
Triceratops, 16,35,55,81
Trochodendroides, 134
Troodon, **11**
Troodontids, 11
Trytilodon, 63
Tsagan Khushu, excavations, 78,108
Tsintaosaurus, 15,36,86
Tuatara, 18
Tugrikiin Shireh, 124
Tumanova TA, 123
Tuojiangosaurus, **15**,37
Tylocephale gilmorei, 116,117,118
osteology, **117**,118
Tylosaurus, **6**
Tyrannosaurus, 12,80,82
Ulan Bator, 43,68
Urbanek A, 113
Urga, see Ulan Bator
Valley of the Big Lakes, 96
Varanus komodoensis, 19
V. varius, nesting habits, 48
Velociraptor, 11,**12**,23,37,55,56,59,60,61,62,120,127
V. mongoliensis, 60,**61**
osteology, **59**,60
Vertebrate family tree, 8
Vultur monarchus, 45
Walker AD, 22
Wellnhofer P, 21,128
Whetstone KN, 22
Williston SW, 23
Wolf Camp, excavations, 64
Woodward J, 30
Xanthos of Sardes, 27
Xenophanes of Colophon, 27
Yaverlandia, 114
Young CC, see Zhong-jian Y
Yunnan Province, 64
Yunnanosaurus, 13
Zalambdalestes, 63
Zhong-jian Y, 36
Zigongosaurus, 102